WEAPON

THE PATTERN 1853 ENFIELD RIFLE

PETER SMITHURST

Series Editor Martin Pegler

First published in Great Britain in 2011 by Osprey Publishing,
Midland House, West Way, Botley, Oxford, OX2 0PH, UK
44-02 23rd Street, Suite 219, Long Island City, NY 11101, USA

E-mail: info@ospreypublishing.com

© 2011 Osprey Publishing Ltd.

All rights reserved. Apart from any fair dealing for the purpose of private study, research, criticism or review, as permitted under the Copyright, Designs and Patents Act, 1988, no part of this publication may be reproduced, stored in a retrieval system, or transmitted in any form or by any means, electronic, electrical, chemical, mechanical, optical, photocopying, recording or otherwise, without the prior written permission of the copyright owner. Inquiries should be addressed to the Publishers.

A CIP catalogue record for this book is available from the British Library

Print ISBN: 978 1 84908 485 7

PDF ebook ISBN: 978 1 84908 486 4

Page layout by Mark Holt

Battlescene artwork by Peter Dennis

Index by Alan Thatcher

Typeset in Sabon and Univers

Originated by PDQ Media

Printed in China through Worldprint

11 12 13 14 15   10 9 8 7 6 5 4 3 2 1

Osprey Publishing is supporting the Woodland Trust, the UK's leading woodland conservation charity, by funding the dedication of trees

Conversions
The measurements in this book are provided in imperial only. For conversion to the metric equivalents note:

miles to kilometres = multiply miles 1.609
yards to metres = multiply yards 0.914
feet to metres = multiply feet 0.306
inches to centimetres/millimetres = multiply inches 2.54/25.4
tons (UK) to tonnes = multiply tons 1.016
pounds to kilograms = multiply pounds 0.454
ounces to grams = multiply ounces 28.350
grains to grams = multiply grains 0.065
drams to grams = multiply drams 1.77

Acknowledgments
No book of a factual nature can ever be one person's work, and a simple testimony of that is the citing of references – any work of a historical nature has to rely heavily upon the work of others in the past. Nor can a book of this nature be 'the last word' on the subject. That would require infinite knowledge and wisdom and access to facts not yet rediscovered. This book, in attempting to give an insight into the conception, birth and relatively short life of the Enfield Rifled Musket, Pattern 1853, is, then, the distillation of the reported facts which have been encountered. As Cardinal Newman said, 'A man would do nothing, if he waited until he could do it so well that no-one would find fault with what he has done.' I therefore accept full responsibility for any errors, opinions, misinterpretation or faulty inferences drawn from those facts.

Assistance was not sought only from the past. In the preparation of this book I have relied heavily upon the help and expertise of many people of the present and my gratitude is due to the Trustees of the Royal Armouries for permission to use illustrations of items from the collections; and to friends and colleagues at the Royal Armouries, especially Graeme Rimer, Academic Director, Alison Watson, Curatorial Manager, Stuart Ivinson and Chris Streek in the library, who have all been unstinting in providing help and support. Likewise thanks are due to colleagues in other organizations for making available various images, especially Juliet McConnell, formerly of the National Army Museum, Lynda Powell of the Green Howards Museum and Liz Bregazzi of the Durham County Records Office, and their respective trustees. A special thanks to Dean Nelson, Curator of the State Museum in Hartford, Connecticut, USA, for providing access to his own notes and knowledge regarding the use of this rifle in the American Civil War and for supplying the photographs of David J. Naumec (Military Historian, ABPP, Mashantucket Pequot Museum & Research Center, Mashantucket, CT) who generously demonstrated the loading and firing processes used with the Pattern 1853; to Thomas (Tom) J. Stelma of Montgomery, Texas, for making available his knowledge on the Enfield ammunition used in the Civil War, along with specimens of Enfield bullets recovered from, I hasten to add, non-protected Civil War battlefield sites; and last but by no means least, to Gary Ombler for his painstaking photography of items in the Royal Armouries' collections.

Artist's note
Readers may care to note that the original paintings from which the colour plates in this book were prepared are available for private sale. All reproduction copyright whatsoever is retained by the Publishers. All inquiries should be addressed to:

Peter Dennis, 'Fieldhead', The Park, Mansfield, Nottinghamshire NG18 2AT, UK

The Publishers regret that they can enter into no correspondence upon this matter.

Front cover images are courtesy Royal Armouries (top) and the National Army Museum (bottom).

www.ospreypublishing.com

CONTENTS

INTRODUCTION	4
DEVELOPMENT	9
The rifling revolution	
USE	24
Sebastopol to Shiloh: the Enfield at war	
IMPACT	50
Accuracy, industry and surgery	
ACCESSORIES	62
FIRING THE ENFIELD PATTERN 1853	67
CONCLUSION	75
GLOSSARY	77
BIBLIOGRAPHY	78
INDEX	80

INTRODUCTION

The Enfield Pattern 1853 rifle and many of its attributes were so radically different from what had preceded it that it marks a major turning point in British military technology and outlook. However, rifle technology already had a long history by the time the Pattern 1853 was introduced.

The concept first appeared in the era of the wheellock. This is not surprising since the wheellock itself was a major innovation in firearms technology, its mechanism providing the first self-contained powder-ignition system. In the wheellock, a hardened serrated steel wheel was wound up against a powerful spring. When loaded, a piece of pyrites held in the jaws of a 'dog' was held against the edge of this wheel, protruding through a slot in the bottom of the priming pan adjacent to the vent in the barrel and containing a small quantity of gunpowder. Pulling the trigger released the wheel which, driven by the spring, rotated rapidly, throwing off a shower of sparks from the pyrites and igniting the powder in the pan. The flame from the burning priming powder passed through the vent and ignited the main charge in the barrel. The wheellock combined features of the technology of locks and clocks and its origin is credited to Nuremberg, a town with a European reputation for lock and clock making at that time.

It has been suggested that a barrel with grooves inside it was first used around 1567, though these grooves were straight, simply to ease loading by providing suitable crevices for the gunpowder residue, or 'fouling', to collect in. The same author suggests that shortly after this a gunmaker named Koster in Nuremberg cut barrel grooves in a long, slow spiral rather than a straight line and rifling as we know it was born.[1] The fact that a projectile spinning on the axis of its line of flight had greater stability began to be understood and appreciated. The projectile was made a tight fit in the bore so that when the explosion occurred, the projectile

---

[1] Busk, H., *The Rifle and How to Use It*

Riflemen open fire during the Second Battle of Corinth, Mississippi, 1862. Originally developed for the British Army, the Pattern 1853 was the first rifled longarm to be widely issued to line infantry. It heralded a revolution in the firepower of the common soldiery and was subsequently used on both sides in the American Civil War; particularly by the Confederacy, who probably used at least some Enfields in every engagement. (Mary Evans Picture Library)

accelerated along the bore and engaged with these grooves which caused it to spin rapidly.[2] It acted like a gyroscope and the same principle is employed today in such devices as automatic pilots and ships' stabilizers.

The numbers of fine wheellock rifles surviving are testament to the existence of and the value placed upon the rifle, as opposed to the smoothbore gun, at that time. The cost of making the complex wheellock limited it to being the chosen firearm of the wealthy sportsman. The poor soldier had to make do with his crude matchlock. The matchlock was essentially the oldest and simplest of firearm technologies, in which ignition was achieved when a length of smouldering twine or matchcord (held either in the hand or in a 'mechanical hand' or serpentine attached to the gun) was thrust into a pan of priming powder.

By the middle years of the 17th century, a new mechanism, the flintlock, had appeared. Like the wheellock, it generated a spark at the instant of firing but did this by striking a piece of flint against a piece of hardened steel; it thus contained fewer components, and so was simpler to make and handle. It was cheaper to make than the wheellock but more sophisticated and effective than the matchlock, and its cost enabled flintlocks to become

---

[2] Since the rifling in the Pattern 1853 made one full turn in 6ft 6in, and assuming the bullet travelled at 1,000fps, this would give a rotational speed of just over 9,000rpm. This is not fast by today's standards but because it was a heavy projectile it still acquired gyroscopic stability.

The Baker rifle, perhaps more than any other of its era, has become widely known to non-firearms enthusiasts through the widespread publicity given to it by the *Sharpe* series on television. (Royal Armouries XII.148)

the standard military firearm. It also eliminated the risks associated with having smouldering matchcord in the vicinity of loose powder, or having the smouldering match doused in severely wet weather. The 18th century became the great era of the flintlock musket, typified in England by the so-called Brown Bess and her numerous offspring.

The popularity of the rifle gradually increased through the flintlock era, but while flintlock rifles were used by the hunting fraternity, the skill needed to make full use of these weapons' potential was beyond the ordinary infantryman. However, in the late 18th century a rifle designed by Ezekiel Baker was adopted in Britain for use by a specially trained regiment of marksmen, the 95th Rifle Regiment.

The great problem with any muzzle-loading rifle was the loading of a tightly fitting ball. With a musket, the lead ball was a tolerably loose fit – it would almost drop down the barrel. The negative side to this was that there was a small gap between the ball and the inside of the barrel, known as windage, which allowed some of the explosion gases to escape around the edges of the ball. The ball was often wrapped in paper or cloth to reduce this dissipation of the explosive power, which would otherwise result in a loss of velocity and, therefore, of range.

On the other hand, the loading of a ball that fitted the bore tightly enough that it would engage with the rifling when fired had its own problems. Usually the ball was wrapped in a greased paper or thin linen patch which could be forced into the rifling grooves. However, gunpowder leaves solid residues in the bore after firing and the rifling grooves provided ideal crevices in which this residue could lodge. What was already hard work became even harder and riflemen were issued with mallets to help drive the ball down the bore after several shots had been fired.

In an attempt to overcome this problem, methods of loading from the breech end of the barrel were developed, some with more success than others. In the wheellock era breech-loading guns were not unknown and often utilized removable

Detail of a breech-loading wheellock gun showing the hinged block in the open position ready to receive its pre-loaded chamber. (Royal Armouries XII.11127)

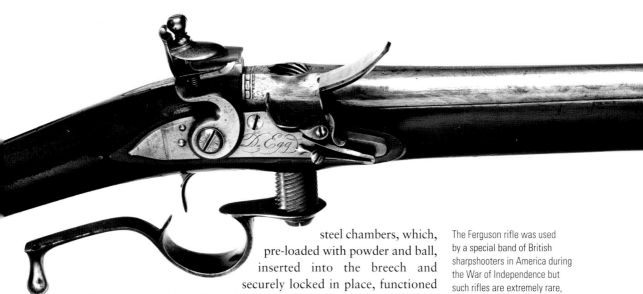

The Ferguson rifle was used by a special band of British sharpshooters in America during the War of Independence but such rifles are extremely rare, especially in Britain. This example is Patrick Ferguson's own which he left behind with his brother when he sailed for America. (Royal Armouries XII.11209)

steel chambers, which, pre-loaded with powder and ball, inserted into the breech and securely locked in place, functioned as rudimentary cartridges. Such chambers had to be a reasonably close fit to minimize the escape of the explosion gases around the edges, and for leisurely sporting use the use of pre-loaded chambers provided a practical solution. Not so for the military which could not afford wheellock guns.

The alternative was to adopt, in place of removable chambers, a breech-loading principle which used loose powder and ball, just as in muzzle-loaders. In Britain the design of Patrick Ferguson was a classic. With the muzzle pointing downwards, turning the trigger guard unscrewed a plug, exposing the breech end of the bore which could then have a ball inserted, followed by the charge of powder. Because the ball was very slightly larger than the bore, it did not just roll out and drop on the ground! Turning the trigger guard in the opposite direction closed the breech once more. It was a system that worked but was beset by the same problems as before. The residues from the gunpowder now clogged the screw threads, making closing the breech increasingly difficult as more shots were fired.

The problem was not really solved until the first few decades of the 19th century had passed and the percussion system was beginning to replace the flintlock mechanism. This in itself was a major improvement. Instead of having to create a spark from a piece of flint striking a piece of steel, chemistry was being applied by that ardent shooter, amateur chemist and mechanic, the Rev Alexander Forsyth. The compounds gold and mercury fulminate are extremely sensitive. Striking a grain of either chemical causes it to detonate. Whereas a propellant explosion occurs over a very short time, a fulminate detonation is virtually instantaneous, and fulminate is far more powerful than gunpowder. Eventually a method was perfected whereby a very small amount of fulminate is placed in a small copper container or cap and this is then placed on a hollow pillar, or nipple, fitted to the breech

The principle of percussion ignition, illustrated by a sectioned portion of a loaded Enfield Pattern 1853 barrel. The hollow nipple is screwed into a projection on the barrel at the breech and a cap is shown in place upon it. When struck by the hammer, the fulminate in the cap detonates and the flame follows the path highlighted in red, into the charge of powder behind the bullet. (Private collection)

ABOVE
A sectioned Brunswick rifle barrel, showing the two helical grooves and the two notches at the muzzle to aid alignment during loading. (Royal Armouries XII.)

BELOW
A diagram showing alignment of the belt on the ball with the notches and rifling grooves. (Private collection)

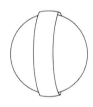

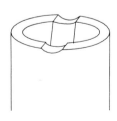

of a gun, and the cap is then struck. The flash from the resulting detonation passes down the hollow stem and ignites the main charge in the barrel. This was a far simpler and more reliable system than the flintlock. It did not, however, solve the problem of loading a rifle from the muzzle.

To overcome having to force a tightly fitting ball down the bore, some attempts were made with projectiles that had projections on their exterior which matched the grooves in the barrel. A classic of this system used by the British military is the Brunswick rifle, which had two deep rifling grooves and fired a ball with a raised belt around it. The idea was that the belt was aligned with the grooves and then forced down the barrel.

But even this did not overcome the problem created by fouling and in 1852 it was reported that 'The loading of this rifle is so difficult that it is wonderful how the rifle regiments can have continued to use it so long. The force required to ram down the ball being so great as to render a man's hand much too unsteady for accurate shooting'.[3]

It can be imagined that in the heat of battle when hands were not too steady anyway, the loading of such a rifle presented additional problems. However, in 1844 Captain Claude Etienne Minié of the French Chasseurs provided an elegantly simple solution to this problem.

It was based on an idea of Captain Henri-Gustave Delvigne of the French infantry who, in 1842, had patented a lead bullet with a cavity in its base, the idea being that the high-pressure gases resulting from the explosion of the propellant would force the edges of the bullet surrounding this cavity outwards into the rifling grooves. Minié took this one step further by deepening the cavity and inserting a conical iron plug. This projectile fitted freely into the bore and, when fired, the force of the explosion theoretically drove the plug forward like a tapered piston and caused the base of the projectile to expand and engage the rifling grooves.[4]

While some, such as John Jacob and Joseph Whitworth, persisted with special bullets shaped to match the geometry of the bore, it was the Minié principle which was to find worldwide acceptance by most military and sporting shooters. In 1851, incorporating Minié's idea, the first step was taken towards the British military establishment gaining the Enfield Rifle Musket, Pattern 1853, one of the finest military firearms of its time.

[3] *Report of Experiments with Small Arms carried on at The Royal Manufactory, Enfield*
[4] In fact, without wishing to detract from Minié's contribution, the idea was not entirely new. It is recorded in the Proceedings of the Ordnance Select Committee in 1857 that the previous year William Greener, the well-known gunmaker, had been awarded £1,000 in recognition of his idea for the use of expanding bullets, proposed to the Board of Ordnance in 1836.

DEVELOPMENT
The rifling revolution

Prompted by what was happening in Europe, especially France, in the second quarter of the 19th century regarding the introduction of rifles as the standard infantry weapon, even the Iron Duke as commander-in-chief had to acknowledge the need for improvement. The view that 'what was good enough at Waterloo is good enough now' could no longer be accepted. Yet even during the Napoleonic Wars, the evidence shows that the musket was hardly good enough. We are told by Sir James Emerson Tennent, a former artillerist, that at the battle of Salamanca, for instance, only 8,000 men were killed or wounded, even though three and half million cartridges had been fired. In other words, only one shot in 437 had any effect! The same source quotes an officer engaged at Waterloo saying that:

> ... he could not see more than three or four saddles emptied by the fire of one side of a square of British infantry upon a body of French cavalry close to them; yet Bonaparte complimented our men on the superior steadiness of their aim. During the Continental campaigns he and his marshals held that 450 yards was a safe distance from small arms, the rifle included.[5]

Hans Busk, scholar, soldier and author, tells us[6] that another commentator, Colonel Schlimmbach of the Prussian artillery, gave it as an indisputable fact that, on average, a man's own weight in lead and ten times his weight in iron were consumed for every one man placed *hors de combat*. He also goes on to tell us that even as late as 1851, British soldiers at the Cape, armed with

[5] Tennent, Sir J. Emerson, *The Story of the Guns*
[6] Busk, H., *The Rifle and How to Use It*

At the siege of Sebastopol, men of the Rifle Brigade nonchalantly snipe at the Russian gunners from a 'rifle pit' while some hot food is being prepared. The Crimean War saw the first issue of the new rifled musket. (*Illustrated London News*, December 1854/Private collection)

smoothbore muskets, fired 80,000 rounds with the result of killing or disabling only 25 of the enemy. Napoleon's inspector-general of artillery, General Gassendi, estimated that 3,000 cartridges were needed for every man disabled; Decker estimated as many as 10,000! There is even an anecdote about soldiers being instructed, in firing at a man at 600 yards, to aim 130ft above him – or in other words, 'if you wish to hit a church door, aim at the weather-cock!'

This obvious inadequacy of the smoothbore musket was certainly supported by trials carried out with the Pattern 1842 musket at Chatham in 1846 by Captain McKerlie of the Royal Engineers, whose comments leave little room for doubt:

> ... as a general rule musketry fire should never be opened beyond 150 yards, and certainly not exceeding 200 yards; at this distance half the number of shots missed the target, 11ft 6ins. [and 3ft wide], and at 150 yards a very large proportion also missed; at 75 and 100 yards every shot struck the target, 2 ft. wide; and had the deviation increased simply as the distance, every shot ought to have struck the target 6 feet wide at 200 yards; instead of this, however, some were observed to pass several yards to the right or left, some to fall 30 yards short, and others to pass as much beyond, and the deviation increased in a still greater degree as the range increased. It is only then under peculiar circumstances, such as when it may be desirable to bring a fire on field artillery, when there are no other means of replying to it, that it ought ever to be thought of using the musket at such distances as 400 yards.[7]

[7] Bond, H., Lt-Col, *Treatise on Military Small Arms and Ammunition*

Yet the value of the rifle over the smoothbore musket should have been eminently apparent after Britain's problems in America in 1776 and again in 1812, despite Bonaparte's view. To some, this value was apparent but their views were not heeded.

However, in 1851 the British Army entered upon what was to become a revolutionary change when the Marquis of Anglesey, Master-General of Ordnance, established the Minié rifle as a replacement for the smoothbore musket as the standard issue to all infantry and 28,000 were initially ordered into production. This 'Regulation Minié Musket' had an overall weight, with its bayonet, of 10lb 8¾oz. It was fitted with a 39in barrel weighing 4lb 10oz with a bore of 0.702in, rifled with four grooves having one turn in 78in. It used a charge of 2.5 drams of powder and fired a 680-grain (1.55oz) bullet 1.03in long with a diameter of 0.69in.[8]

The 'Minié' was not so much a new design. In effect it simply followed the design of muskets at that time, with little visible change apart from its barrel, so that at first glance it could be taken for a common musket of the early 19th century. The only give-away was the long rear ladder sight. It was never generally issued although it did see service in the Eighth Xhosa War of 1851 and then again in the Crimea at the battles of Alma and Inkerman. Shortly after the Minié was brought into service, large numbers of the Pattern 1842 musket were converted into rifles and issued for use by the Royal Marines, hence the term 'sea service rifle'. These were of even larger calibre, with a bullet weighing 825 grains and having a diameter of 0.731in, and using a charge of 3 drams of powder.

The bullet used in the Minié was the original design of totally conoidal form. As it had no cylindrical portion, and hence no parallel sides, it was found impossible to ensure proper alignment in the bore, with the result that the bullet often left the bore slightly canted to one side, leading to inaccuracy. To overcome this, a new design was introduced which was cylindro-conoidal, that is to say, behind the conoidal front portion it had a cylindrical portion with parallel sides and this ensured that, in loading, the axis of the bullet coincided with the axis of the bore. It had the same weight as its predecessor but to compensate for the extra metal, was ten thousandths of an inch shorter and was fitted with an almost hemispherical iron cup in its larger hollow base.

It quickly became obvious that the Minié rifle was not the ideal weapon and in 1852 the new Master-General of Ordnance, Viscount Hardinge, ordered that experiments be carried out to determine the best design of rifle, combining lightness and efficiency, for military service.

The Regulation Pattern 1851 Minié rifle, although a major technological step forwards, differed little in appearance from the Pattern 1842 musket, retaining features such as the attachment of the barrel to the stock by pins and the use of ramrod pipes. (Royal Armouries XII.1907)

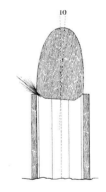

This shows the result of having a misaligned conoidal Minié bullet of the original pattern. (Royal Armouries, 1852 *Report of Experiments with Small Arms*)

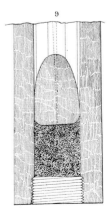

---

[8] Note the dram and grain weights referred to are Avoirdupois measure: (1 grain = 0.065 grams; 27.34 grains = 1 dram; 1 dram = 1.77 grams; 16 drams = 1 ounce = 28.34 grams)

Regulation Minié

Cylindrical Minié

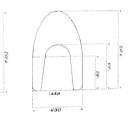

The old (top) and new (above) patterns of Minié bullet. (Royal Armouries, 1852 *Report*)

In the course of these experiments, detailed in a report of 1852, various rifles submitted by the private gunmakers Westley Richards, Greener, Wilkinson, Purdey and Lancaster, as well as those submitted by Lovell on behalf of the government, were tested. One interesting observation made during the course of these trials relates to the concern expressed by some regarding the possible reduction of bore size:

> ... and it is maintained by the advocates for a large bore, that the wound from a bullet of this diameter is more dangerous than from a bullet of the smaller bores, in proof of which it is said that wild animals are found to run further when wounded with a small ball than they do when wounded with a large one; but this reasoning does not seem applicable to the human race, for it is presumed that few men would be found willing to move far when wounded by a musket ball, whether the hole in their body was .702 or .530 of an inch in diameter.[9]

The outcome of these trials was the creation of two rifled muskets at Enfield which embodied the improvements and alterations suggested by the rifles submitted and which, it was hoped, would form the basis of the new military rifles. These two rifles had 39in barrels of 0.577in bore, uniformly rifled with three grooves undergoing one half turn in the length of the barrel – in comparison with the bullet used in the Pattern 1842 sea service rifle, this was certainly 'small bore'.

In the new rifle, the barrel was fastened to the stock by three iron bands which also served to retain the ramrod in its channel and with their use came the disappearance of brass ramrod pipes; the ramrod had a swell near the head to prevent the hand from sliding down the shaft; the lock was highly refined compared with what had gone before and was fitted with a swivel, which reduced friction compared with the old 'hook' locks; and the bayonets were secured by means of the French-pattern locking ring.

The Pattern 1842 rifle bullet on the left compared with the 1853 bullet on the right gives an indication of the reduction in bore size which had taken place in the development of the new rifled musket. (Private collection)

[9] *Report of Experiments with Small Arms carried on at The Royal Manufactory, Enfield*

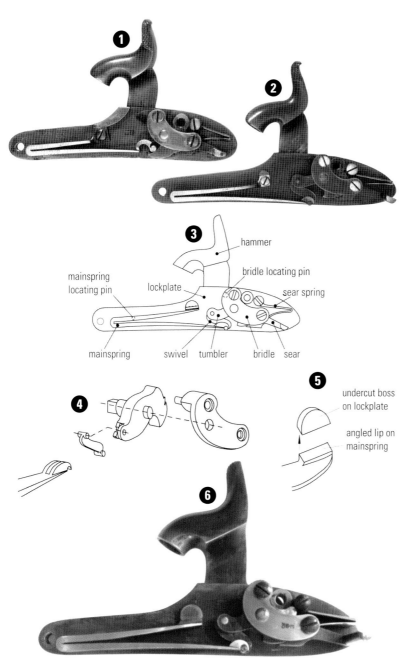

1. The old style 'hook' lock as used on the Minié rifle. The working tip of the mainspring was in the form of a hook which acted on an extension of the tumbler. Because it 'rubbed' on this extension when the lock was in action, friction was created, slowing the action of the lock and reducing the force used to strike the cap. (Royal Armouries XII.1907)

2. The transitional 'swivel' lock used in the first experimental Pattern 1853 rifles, in which a swinging link was used to connect the end of the mainspring with the tumbler. Because of the limited points of contact, friction was minimized. Both locks use a screw to secure the mainspring to the lockplate. (Royal Armouries XII.2981)

3. The components of the Enfield Pattern 1853 lock. (Private collection)

4. Schematic exploded view of bridle, tumbler, swivel and the claw on the mainspring which engages with the swivel. (Private collection)

5. Diagram showing how the end of the mainspring lodged under the boss on the lockplate, thereby securing it in place. (Private collection)

6. In the final version of the lock for the Pattern 1853 rifle, the spring retaining screw was replaced by a simple undercut block into which the corresponding bevelled tip of the mainspring slid and was retained in place. It made assembling and dismantling simpler and it meant there was one less screw to manufacture or be lost in the field. (Royal Armouries XII.1918)

One of these first two rifles was made with a block rear sight, with an additional two leaves for 100 yards and 200 yards range. It was felt that this simple sighting arrangement was appropriate for troops of the line, militia, etc. The second rifle had the same block sight but, in addition, was fitted with an adjustable 'ladder' sight, based on those designed and patented by Westley Richards, and graduated to allow aimed shots at ranges up to 800 yards to be made. This rifle, with its more refined sight, was felt more appropriate for equipping rifle regiments and special corps. There was still, evidently, some reluctance to create total equality!

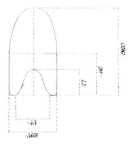

Pritchett's bullet, like the 'official' bullet, was cylindro-conoidal, but relied simply on the force of the explosion gases to expand the skirt of the hollow base into the rifling grooves. (Royal Armouries, 1852 *Report*)

For ammunition, Pritchett, the London gunmaker, was asked to adapt a projectile he had been successfully experimenting with, which simply had a hollow base and dispensed with the cup or plug.

These rifles, at distances of up to 800 yards, were found to shoot as accurately as any that had so far been tried. Shooting was carried out with paper cartridges made up in the 'usual Minié style', that is with greased bullets which had to be reversed for loading. The alternative method (as employed with the smoothbore muskets using spherical ball) in which the bullet was loaded without reversing, was also tried. For those not familiar with these terms, it simply applies to the alignment of the bullet within the cartridge, the 'reversed' bullet having its nose pointing inwards (see page 67). It was found that in the case of the 'non-reversing' cartridge, fouling occurred immediately above the powder for about 2in. In the Minié system with the bullet reversed, the greased paper around the bullet lay immediately above the powder charge, ensuring that the entire barrel was lubricated and wiped clean with every shot.

After further refinement of these experimental weapons, the Pattern 1853 Rifled Musket theoretically came into existence. However the idea of having two different sighting arrangements, one for 'ordinary' soldiers and one for rifle regiments, seemed to defeat the whole idea of the exercise and in 1854 the sighting arrangement was changed, the resulting rifle receiving its distinctive ladder rear sight. This was, in fact, a combination of Westley Richards', Lancaster's and the Ordnance pattern used on the Minié. In Westley Richards' sight, the leaf could be folded flat forward or backward; Lancaster's sight bed had protective wings at the side and in the Ordnance pattern, the leaf was held in the vertical position by a spring. Usually a 'sealed pattern' (see page 16) would have been created to act as the reference standard to guide manufacture and inspection but no such sealed pattern is recorded in the Enfield Register. The earliest true Pattern 1853 rifle in the Royal Armouries collections is that shown opposite. The blending of various features of the rifles submitted for these experiments into this production rifle show that the experiments were successful and that those conducting them were able to assess the features which were likely to be of value.

Its overall conception and design made the Enfield Pattern 1853 rifle, rather than its predecessor the Minié, a major turning point in British military thinking. This improvement was apparent in various ways. It indicated a complete shift away from the design ethos of the 'Brown Bess' of the 18th century, to which even the new percussion muskets of the first few decades of the 19th century had shown a family resemblance. Most noticeable perhaps was the slimmer, more streamlined appearance, which can be accounted for by the dramatic reduction in bore size from 0.702in to 0.577in and the consequent reduction in barrel diameter. Detailed changes in the shaping of the woodwork are also noticeable, especially in the fore-end which is progressively slimmed down in a series of steps. One of the effects of these changes was a 12 per cent reduction in weight from that of the Minié, to 9lb 3oz. What makes it really stand out, however, is the use of the iron bands to retain the barrel, the method used by the French, for instance, from the middle years of the 18th century.

Another French influence on the Pattern 1853 can be found in the bayonet. British socket bayonets had relied upon the 'zig-zag' slot to fix them to the muzzle, sometimes using the block foresight but often employing a 'bayonet stud' on the underside of the barrel. For a straight thrust and pulling back, this was entirely adequate. If, however, the musket was turned in the wrong direction the bayonet could be pulled off, which was not a desirable occurrence in a bayonet engagement. This was of course recognized and a variety of methods of securing the bayonet were tried – some experimentally, others more extensively, but all left something to be desired. On the Minié rifle, for instance, a small projection from the stock beneath the muzzle can be seen. This is known as Lovell's spring catch, and was designed to engage with a step on the collar of the bayonet, thereby preventing its accidental removal; it can be found in use with the 1839, 1842 and 1851 Pattern arms.

While the fixing of the bayonets with Lovell's spring catch was fairly straightforward, removal was another matter. Anyone who has attempted to operate Lovell's catch will know it has a powerful spring and requires considerable force to move it sufficiently to disengage the bayonet. This is not made easier by the smooth finish to the 'thumb pad' on the tip of the spring, and is even worse when the thumb is either wet with rain or sweat, or, after firing, greasy from holding cartridges with lubricated bullets.

The French had in fact overcome the problem as early as 1777 by fitting a partially rotating locking ring on the socket of the bayonet. With the ring in the correct position, the bayonet could be easily fitted and with a slight turn of the ring the bayonet was secured in such a way that it could not be removed accidentally. The Pattern 1853 incorporated identical mechanical features.

One source has claimed that the Enfield 1853 rifle was 'patterned after' the American Springfield. Such a claim defies comprehension, if only because the Springfield infantry rifle was not developed until 1855! The various US military muskets preceding it owed their appearance more to the French musket of 1777 which so delighted Jefferson and, if anything,

An early example of the first model, Pattern 1853 Rifled Musket, dated Tower 1854 and incorporating the first pattern ramrod. (Royal Armouries XII. 3064)

HOW LOVELL'S CATCH WORKED

LEFT
The bayonet was slid down the barrel, rotating it to allow the foresight to follow the 'zig-zag' slot, so that the projecting lug sat on the face of the thumb piece on the locking spring.

RIGHT
The bayonet was then pushed home, the lug riding over the locking spring which then snapped back into place, securing the bayonet on the muzzle and preventing its removal until the catch was manually disengaged. (Royal Armouries XII.366 and X.107)

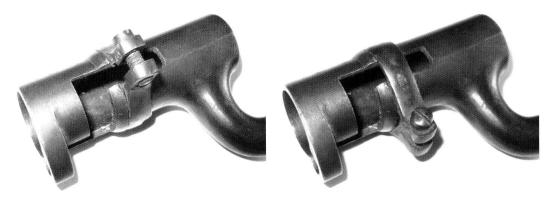

Socket of the Pattern 1853 bayonet showing (left), the locking ring in position to allow the foresight to pass into the end of the slot and (right) the ring rotated to engage with the rear of the foresight block and hold the bayonet in place. The ring has a ramped edge so that it can be made a tight fit against the foresight. (Private collection)

the Springfield 1855 rifle owed something to the Pattern 1853 in its appearance, with the abandonment of the French style of furniture and the use of the simple brass fore-end tip which is to be found on the Enfield.

As a result of the various delays and modifications to things like the sighting arrangement, the Pattern 1853 in its familiar form was eventually introduced into service in 1855 and, for the first time in reality, a weapon was placed in the hands of *every* soldier which far surpassed the old smoothbore musket in every conceivable respect.

Once the pattern of rifle had been established, it was subjected to a more or less continual process of evaluation and improvement, as well as forming the basis for a number of firearms derived from it. In all, there were around 15 direct derivatives for British service; added to these were the Lancaster and Whitworth rifles, plus those specially created for use in India after the Mutiny, creating around 30 in total. This great variety makes it a rich if sometimes confusing area for both collectors and historians, but means that in the space available here it will not be possible to study each one in detail.

At the time when the Pattern 1853 rifle was introduced, all arms that were approved for service had, theoretically at least, a 'sealed pattern' created. It was known as a sealed pattern because, when the final design was agreed, a sample was made and the red wax seal of the Board of Ordnance was attached to it, either directly or by means of a securely affixed label. Such patterns served the purpose of being *the* reference standard to guide both manufacture and inspection to make sure any weapons of a particular pattern conformed to specification. It was a system first introduced by Charles I who was dismayed at the great variety amongst ostensibly 'standard' weapons and related equipment, and who instructed that a greater degree of uniformity amongst weapon types be introduced. The sealed patterns were there to ensure adherence to approved designs, serving as the last court of appeal when it came to any dispute. Today they serve a similar function, not for manufacturers but for collectors and historians. Not surprisingly, sealed patterns are often unique and while on occasion several might have been created to issue to contractors, they remain extremely rare.

However, a sealed pattern was not created for every small change and although the Pattern 1853 rifle underwent a variety of modifications and improvements throughout its service life, it retained its designation as Pattern 1853. In this it differed from the derivatives, whose designations changed.

THE RIFLE MUSKETS

For whatever reason, a sealed pattern of the first 1853 Pattern rifle musket does not seem to have been prepared, or if one was prepared, its whereabouts, if it survives, is unknown. In 1859 a modification was approved, consisting of shortening the butt by one inch; that is to say, the distance from heel plate to trigger guard was reduced from 14in to 13in. The change involved is not readily apparent but was undertaken to allow the rifle to be accommodated more easily, and therefore used more effectively, by soldiers of lesser stature than those in, say, the Guards. This modification did warrant the creation of a sealed pattern but did not incur a change in title; the new rifle was still the Pattern 1853 and long and short butt versions were issued side by side, usually in a ratio of one-third long and two-thirds short butts, unless experience or demand dictated otherwise. How far the private trade might have followed this course in supplying rifles to other customers is a question open to speculation.

The first rifles had what were termed 'screw bands' securing the barrel to the fore-end of the stock. Whilst these performed that function well, the projecting heads and tails of the screws were inclined to snag and damage uniforms. The result was that a second pattern was introduced in which solid bands were used. These had a totally smooth contour and were unlikely to snag or damage anything; they were retained on the stock by a simple spring catch arrangement. In theory at least they could easily be removed by pressing the spring into its cavity in the stock, which then allowed the band to slide over it, and they are often referred to by today's collectors as 'spring bands'. Most, if not all, of the rifles manufactured under contract in Liège have this feature. These Liège-made rifles are also distinguished by having the date on the lockplate italicized. Similarly, those made under contract by Robbins and Lawrence at Windsor, Vermont, are also fitted with 'spring bands' but are unique in carrying the name 'Windsor' on the lockplate, instead of the usual 'Tower'.

However, the springs and bands were found to become either tight or loose, depending on the climate, so in 1861 another new form of screw band was introduced as a replacement for the bottom and middle bands of earlier pattern. In this modified type, often referred to as the 'Baddeley band', the band was thickened close to the joint so that the heads and tails of the screw were shrouded, providing a smooth contour and obviating the tendency to snag and damage clothing or accoutrements. This adaptation also led to a saving in manufacturing costs.

Also in 1861, a simple modification of the rearsight was undertaken, consisting of graduating it to 1,000 yards instead of 900 yards as previously. The shoulder of the leaf upon which the sliding bar rested for the 500-yard range was to be lowered by approximately 0.125in and the 'V's on the cap of the leaf were to be cut differently, but we are not told exactly how.

The sealed pattern rifle created in 1859 following a shortening of the butt by one inch. (Royal Armouries XII.1918)

BARREL BANDS

LEFT
First pattern with projecting heads and shanks of the screws. It may be worth noting that in all screw bands, the screws enter from the left when the rifle is held in the firing position. (Royal Armouries XII.1918)

CENTRE
Second pattern of barrel band. The retaining spring can be seen projecting from the stock. (Royal Armouries XII.1365)

RIGHT
Third pattern 'Baddeley' barrel band in which the screw is fully shrouded. (Royal Armouries XII.1928)

Whilst not strictly speaking derivatives, the rifles produced at Enfield after 1857 were very different from their predecessors simply because all the components were standardized to such a degree that they were interchangeable, whereas all earlier components, made by contractors, were not.

THE CARBINES

Artillery carbines

The first derivative to appear was the artillery carbine and, though designated the Pattern 1853, one example exists with an 1852 date; it was created as part of the 1852 series of experiments, and has a curious structural feature, namely the long rib extending from the bayonet bar on the barrel towards the muzzle. Like the rifle, this carbine has the early type of ramrod with mushroom head and a swell on the shaft. The production version differed only slightly, in having a modified bayonet bar with a much shortened rib extension.

In 1858 this was again modified so that the bayonet lug had no extended rib. This feature was continued in a new version of the carbine, the Pattern 1861, which was created following a change in the rifling from three grooves to five grooves and the fitting of a modified rearsight, and using the Baddeley-type barrel band.

Experimental Pattern 1853 artillery carbine, dated 1852. (Royal Armouries PR.5173)

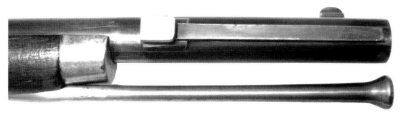

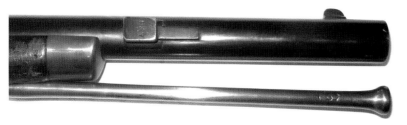

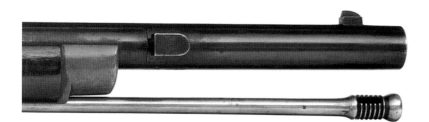

TOP
Muzzle of the experimental carbine showing the long rib extending from the bayonet stud.

CENTRE
Muzzle of the production version, or second model, of the Pattern 1853 artillery carbine showing the different bayonet fitting with its short extension rib.

BOTTOM
Pattern 1861 artillery carbine – note the absence of any forward projecting rib on the bayonet bar, the use of the second pattern ramrod and a change in the shape of the foresight blade. (Royal Armouries XII.1928, PR.5145 and PR.5173)

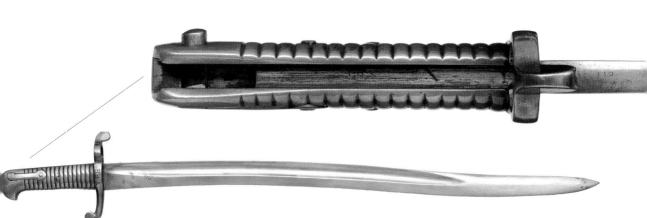

These carbines were designed to take a sword bayonet of distinctive form with a so-called 'yataghan' type of blade with a shallow double curve. The first pattern for the experimental carbine was even more distinctive in having a brass hilt. The two bayonets which succeeded it had steel hilts with chequered leather scales. And, while each looks identical, the nature of the bayonet bars on each of the three carbines, aside from any other possible differences, is reflected in the edge of the hilt which lies against the barrel.

The experimental Pattern 1853 artillery carbine bayonet, with detail of the hilt, showing the long groove to match the rib on the muzzle. (Royal Armouries, ex-MoD Pattern Room Collection, PR.1935)

Hilts of the bayonets for the artillery carbine. The second model Pattern 1853 (top) had a much shorter groove to match the short rib on the muzzle. The Pattern 1858 (bottom) had a plain mortice slot corresponding with the lack of a rib. (Royal Armouries, ex-MoD Pattern Room Collection, PR.1935 and PR.2558)

BELOW
Pattern 1856 cavalry carbine. (Royal Armouries, ex-MoD Pattern Room Collection, PR.5148)

BOTTOM LEFT
Detail of the Pattern 1856 cavalry carbine muzzle, showing the rounded top to the foresight to prevent it from getting snagged in the saddle 'bucket' and the 'captive ramrod'. (Royal Armouries, ex-MoD Pattern Room Collection, PR.5148)

BOTTOM RIGHT
The saddle ring served a similar purpose to the 'captive ramrod'. (Royal Armouries, ex-MoD Pattern Room Collection, PR.5148)

Cavalry carbines

The first cavalry carbine introduced into service was the Pattern 1856. Cavalry carbines follow a similar style to that of the artillery carbines but have major differences resulting from practical considerations which create very noticeable distinguishing features.

One of these is the type of ramrod used, generally referred to as a 'captive ramrod', so arranged that it could be easily used but not dropped and lost. Related to this is the 'saddle ring' attached to the left-hand side of the stock, consisting of a bar with a sliding ring fitted on it; this was for attachment to a lanyard, worn across the shoulders of the cavalryman, so that if dropped, the weapon would not be lost. Thirdly, since cavalry carbines were not intended to be used with a bayonet, there is no means of fitting one.

The Pattern 1861 was created for cavalry carbines to accommodate the changes made in that year for the artillery carbines, and a comparison of the old and new sights is shown opposite.

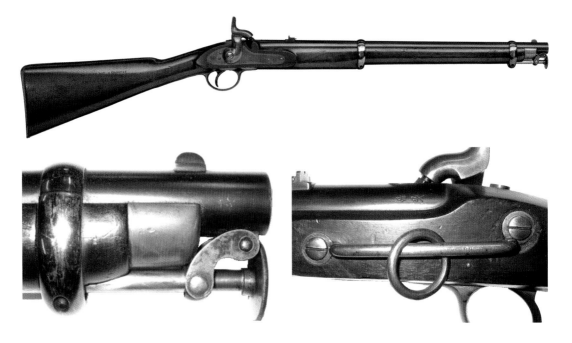

THE SHORT RIFLES

Falling between the rifled musket and the carbines in size were the short rifles, often referred to as the 'two-band' rifles since, being of shorter barrel length than the long rifle, they required only two bands to secure the barrel to the stock. They were felt to be more 'handy' in certain situations such as skirmishing. Various patterns were approved, the most common being the Pattern 1856 which was issued to Sergeants of the Line plus the 60th Regiment, the Cape Mounted Rifles, the Royal Canadian Rifles and the Rifle Corps – which once again made these different from the ordinary line regiments!

The Pattern 1858 naval rifle differed in having a barrel with visibly thicker walls than standard, and it had five-groove rifling. Otherwise it was a close copy of the Pattern 1856. In 1860, because of its superior shooting accuracy compared with the three-grooved short rifles, it was approved to become the standard short rifle for all services and in 1861 it underwent the same changes as all the other arms, being equipped with Baddeley-type barrel bands. One unusual feature of the short rifles, with the exception of the naval rifle, is the use of case-hardened iron furniture.

Detail showing the rearsights of the cavalry carbines, Pattern 1856 (left) and Pattern 1861 (right). These sights were used on both artillery and cavalry carbines. (Royal Armouries, ex-MoD Pattern Room Collection, PR.5148 and PR.5149)

THE INDIAN SERVICE ENFIELDS

Following the Indian Mutiny, there was concern over placing weapons equal to those carried by British soldiers in the hands of the Indian troops. The result was a series of specially created weapons, whose distinction lies in the fact that the majority were smoothbored, not rifled arms. This means of course that they were less accurate, especially as range increased. They were, however, manufactured to the same high standard as their British service counterparts.

Pattern 1856 short rifle. (Royal Armouries; ex-MoD Pattern Room Collection, PR.5148)

Pattern 1858 and Pattern 1859 muskets for Native Infantry

The first of these weapons was the Pattern 1858, which was no doubt prepared by simply boring a standard Enfield barrel oversize to 0.656in diameter. This corresponds very closely to the diameter of the threaded hole where the breech pin is fitted so this standard component could still be used. However, it was felt that these over-bored barrels left too thin a wall, resulting in the barrels being weakened. So, a new Pattern 1859 was created in which this deficiency was overcome and it is this version which, because it was made in much larger numbers, is the most common version found today. Because there was a desire to retain overall similarity in appearance between the rifled and smoothbored arms, they can easily be mistaken at a glance, but more careful scrutiny will reveal on the Patterns of 1858 and 1859 a simple notched rearsight, while on the Pattern 1859 the blade foresight was replaced by what is often referred to a 'dog-kennel' foresight.

These features are also characteristic of all the other firearms in this group and since many of the smoothbore Indian arms are otherwise close copies of the rifled versions, there seems little point in including them all. However, they sometimes did have different purposes, including use by Native Foot and Mounted Police, and have a parallel in the Pattern 1858 issued to the Irish Constabulary which was also smoothbored.

LANCASTER AND WHITWORTH

These two gunmakers suggested alternatives to the typical grooved rifling used in the majority of rifled arms. Lancaster proposed a bore which appears – and is – smooth, but which is not circular in section as in a common smoothbore gun. Instead, the bore is very slightly oval, the major and minor axes differing by only ten thousandths of an inch, and these axes rotate about the axis of the bore in progressing from breech to muzzle. A bullet fired from such a gun is therefore deformed to fit the oval cross-section and rotates as it travels along the bore in the same manner as a bullet in a conventionally rifled barrel. For obvious reasons, this feature is commonly referred to as 'oval bore rifling'. Even during the 1852 experiments, such rifles were found to give results that were equally good as, if not better than, grooved rifling; but it was the Enfield pattern of rifling that was ultimately adopted, mainly on the grounds that to change part-way through production would incur heavy costs and undermine the consistency of standardization being aimed for. However, tests with the Lancaster persisted and in 1855 Lancaster's system was adopted for the production of the Sappers and Miners carbine. In fact, if the possibility of moving to breech-loading had not been imminent, all the Pattern 1853 series might well have become 'Lancasters'!

In size these 'Sappers' carbines are more reminiscent of the short rifle than a carbine, but any chance of confusion is eliminated by the fact that on the Lancaster carbine the rearsight was reversed with reference to the Pattern 1856 short rifle, while, to make things even clearer, the barrel is

The distinctive rearsight (left) and foresight of the smoothbored weapons for Indian service. (Royal Armouries XII.980)

usually stamped 'Lancaster's Patent'. The carbine also had a distinctive brass-hilted bayonet with a so-called 'pipe-back' blade, very different from the usual sword bayonets of the time.

There was little chance of mistaking the Whitworth rifle for a smoothbore! In this instance, the bore was hexagonal in cross-section and, like the oval bore, twisted along the length of the barrel. It was also smaller bore, at 0.451in diameter. Whitworth was convinced that rifling such as he proposed gave far better results than either the oval bore or the more conventional pattern, especially if using mechanically fitting – that is, hexagonal – bullets, and he was to some extent correct. However, the problems of creating special bullets just for use with this rifle were not something that appealed to the military authorities. There was much discussion over the Whitworth and eventually, in 1862, an order was placed for 1,000 of the full length rifle for trials, followed in 1863 by an order for 8,000 short rifles, but they were never used in any engagement. One of the minor distinguishing features of the short rifle, the Pattern 1863, is the bayonet bar on the barrel band closest to the muzzle. In place of the usual rectangular block, it is circular in section and the bayonet has a corresponding circular mortice in the hilt. The Whitworth bayonet and the Enfield-made Whitworth rifles are very rare items, highly prized by collectors.

The Enfield rifle in various forms saw service in a variety of conflicts, notably the Crimean War, the Indian Mutiny and the American Civil War. But it was developed and introduced at a time when another revolution in small arms technology was just beginning to gather momentum – the development of breech-loading, using a self-contained cartridge complete with projectile, propellant and primer. In British military terms the Pattern 1853 was effectively made obsolete on 18 September 1866 with the introduction of the Snider rifle. But because the Snider was a conversion of the Pattern 1853, it utilized nearly all of the parts of the latter with some significant additions. So the production of Pattern 1853 parts continued for a good number of years afterwards until the Snider itself became redundant with the introduction of the Martini Henry rifle in 1874.

USE
Sebastopol to Shiloh: the Enfield at war

THE CRIMEA

No sooner had the Pattern 1853 rifled musket been approved than its first opportunity for testing in action arose. After Turkey rejected Russia's attempts to secure comparable rights with France for the protection of Christian sites and subjects in Ottoman-held Palestine, Russia invaded, occupying Moldavia and Wallachia in the Balkans and refusing demands from other European powers for immediate withdrawal. On 4 October 1853, Turkey declared war on Russia. The Russian invasion of Bulgaria on 20 March 1854 led Britain and France to join Turkey in a declaration of war four days later, and the Crimean War began. It was the first war in which rifles were used extensively by most French and British troops, and was fought against Russian troops who were armed predominantly with smoothbore muskets.

Orders were immediately issued for 1,000 Minié rifles manufactured in 1851 to be sent to Turkey without delay to replace those in use by troops stationed there. From Britain's point of view, the timing of the war could not have been worse. The supply of infantry arms was in a state of flux; the Pattern 1851 Minié rifle had not been manufactured in large enough numbers to equip all the troops, while the manufacture of the new Pattern 1853 rifle was hardly underway. In the meantime it seems that both the 1851 Minié rifle and the Pattern 1853 Enfield rifle were regarded as front-line weapons. If this is so, with their different ammunition, it can only have exacerbated what became an almost chaotic supply chain and placed additional burdens on an already strained commissariat. Detailed information on the issue of the Pattern 1853 rifle is very sparse and so reliance has to be placed largely on accounts of the war carried in

publications such as the *Illustrated London News* and its accompanying illustrations and *The Times*, along with other, more random sources such as personal letters or diaries. Even then, the situation is not helped by the use of the term 'Minié Rifle' in the press when in many cases it is apparent that what is actually meant is the new Pattern 1853 rifle as opposed to the old Pattern 1851 Minié rifle. That such confusion existed in the minds of contemporary commentators is made abundantly clear in an article appearing in the *Illustrated London News* in April 1855 where the 'Government Minié Sight' is illustrated but is, in fact, the totally different Pattern 1853 rifle sight.

While such confusion is understandable amongst journalists, we are left with the problem of trying to unravel what was really meant. Many of the illustrations also lack attention to fine detail, which, although equally understandable, similarly hinders pinning down the use of the Pattern 1853 in the Crimea. In one illustration, for instance, showing the camp of the Rifle Brigade, there is hardly a firearm in sight! However, from the simple analysis of the pictorial evidence we can presume that either or both the Rifle Brigade and Marines were equipped with the new rifle. Also, as this illustration is dated 1855, the year when general issue of the rifle was approved, it is probably safe to assume that the Rifle Brigade would have been one of the first to receive the new weapon.

Gradually, manufacture, procurement and issue of the new British rifle accelerated and it began to reach other regiments too. In a few rare instances photographs do show regiments, such as the 68th Durham Light Infantry, equipped with the Pattern 1853 rifle in the Crimea.

The encampment of the 'Rifles' and Royal Marines on the heights of Balaclava. Although the group in the mid-ground appears to be carrying long arms in some form, these are not identifiable. Only a close examination of the picture (below) reveals that one of the soldiers seated in the foreground is carrying a bayonet on his belt. If we assume that the strange band on the socket is in fact a locking ring, then it can only be the Pattern 1853 bayonet, which would confirm the use of the Enfield rifle. (*Illustrated London News*, March 1855/Private collection)

This photograph of the 68th Durham Light Infantry in the Crimea in 1855 shows them with the new Pattern 1853 rifles. The giveaway detail is the presence of barrel bands – the Pattern 1853 was the first British longarm to adopt these. (Reproduced by permission of the Trustees of the former DLI and Durham County Record Office)

It was in the Crimea that the rifle was first used as an effective long-range weapon, proving powerful against artillery and cavalry alike. The whole objective of the Crimean War was, ultimately, to destroy Russia's power in the Black Sea by crippling and denying use of the great Russian naval base at Sebastopol and undermining any further plans for incursions into the Ottoman Empire. The first engagement en route to this objective was the battle of the Alma. The Russians had constructed the 'Great Redoubt', a heavily fortified position to protect the northern approaches to Sebastopol, on a high ridge overlooking the Alma River. It was an ideal defensive position. Crossing the river was straightforward for the Allies, but the British contingent then faced a hard slog 300ft up a steep hill to reach the redoubt, which was defended by around 36,000 Russians.

The two Light Divisions that made the assault on 20 September 1854 included the 19th, or 1st Yorkshire North Riding Regiment, the 'Green Howards'. Regimental records show that in 1854 they were issued with Minié rifles but even some military commentators have slipped into the same error as journalists in their terminology and so we are left to wonder which rifle was actually issued. A report in the *Illustrated London News* gives some indication of the desperate nature of this assault:

> All accounts agree in representing this bold movement as the grand crisis of the battle of Alma… The Russian general, finding himself completely outflanked, attempted to change his front and drive the French down the hill. Our artillery had been playing upon the enemy for some time, but the time had now come when the bayonet must be used.

This engagement is an important event in the Green Howards' history, providing the unit with its first battle honour, and every year on 20 September the regiment celebrates 'Alma Day'.

Once the Alma had been taken, Sebastopol was in sight and the attack on that stronghold commenced. Meanwhile, Lord Raglan, receiving reports

The Light Division, including the 19th 'Green Howards' Regiment, is shown here cresting the steep approach and storming and capturing the 'Great Redoubt', a heavily fortified Russian earthworks equipped with 14 artillery pieces, at Alma. (© Green Howard Museum Trustees)

of Russian movement across the Tchernaya River towards the Allied supply base at Balaclava, had become concerned about its safety. On 18 September he had visited the Sapoune Ridge, but could see no immediate threat. The Russians, however, under Prince Menshikov, were preparing to advance; they had probed the line of redoubts along the Woronzov (Causeway) Heights and continued to do so in early to mid-October. Eventually, battle commenced at Balaclava on 25 October 1854 when the Russians effectively overran the perimeter defences of the town, including the redoubts housing the Allied artillery. It was in attempting to recover these defences that the Light Brigade charged down the 'valley of death' to their immortal glory in a vain and mistaken attempt to stop the Russians removing the Allied guns.

'Grand Charge of the Guards at Alma'. The Guards also took part in the assault at Alma, and this illustration gives some idea of the terrain involved. Again we are beset with a lack of detail of the soldiers' weapons and only 'generic' longarms are depicted. However, it is extremely likely that the Guards would be one of the first to receive the new rifle. (The Art Archive/Musée de L'Armée Paris/Gianni Dagli Or)

Robert Gibb's *The Thin Red Line* depicts the 93rd Highlanders repelling the charge of the Russian cavalry outside Balaclava. (Courtesy of the Council of the National Army Museum, London)

Detail clearly showing the Pattern 1853 rifles, with bayonets fixed, providing the flash of steel that caught William Russell's imagination, being used by the 93rd. (Courtesy of the Council of the National Army Museum, London)

More successful was the repelling of a Russian cavalry charge by an infantry regiment. Although created some time after the event, Robert Gibb's painting of 1882, *The Thin Red Line*, was inspired by Alexander Kinglake's account of the 93rd (Sutherland) Highlanders at Balaclava and the title was probably derived from war correspondent William Russell's phrase, 'That thin red streak topped with a line of steel' in his account of the event:

> As the Russians came within 600 yards, down goes that line of steel in front, and out rings a rolling volley of Minié musketry.[10] The distance is too great; the Russians are not checked, but still sweep onwards through the smoke… With breathless suspense everyone awaits the bursting of the wave upon the line of Gaelic rock; but ere they come within a hundred and fifty yards, another deadly volley flashes from the levelled rifles, and carries death and terror into the Russians. They wheel about, open files right and left, and fly back faster than they came.[11]

[10] Note again the use of 'Minié' when it is likely the Enfield rifle was being used
[11] Russell, W. H., *The War: From the Landing at Gallipoli to the Death of Lord Raglan*

A Russian cavalry officer is reported as commenting later:

> We did not know you were lying down behind the hill until you started from the ground and fired a volley at us. We were unable to rein up or slacken speed ... before we received your second volley by which time almost every man and horse in our ranks was wounded.

This is in stark contrast to the previously mentioned account of Waterloo where one observer 'could not see more than three or four saddles emptied by the fire of one side of a square of British infantry upon a body of French cavalry close to them'.

As the Russians wheeled aside to escape the Highlanders' fire, they exposed themselves to a third volley into their flank. Asked why he had been so unorthodox as to receive a cavalry charge in line instead of in a square, Sir Colin Campbell is said to have replied, 'I knew the 93rd, and I did not think it worth the trouble of forming a square.' Not only that, but because the regiment was so undermanned, Campbell was forced to draw up his line only two deep instead of the normal three deep, so as to provide a long enough front, and for the men to hold their position in the face of a cavalry charge shows a tremendous and steadfast courage. To repel the charge as well was a just cause for pride and for this action, the 93rd became the only infantry regiment to be awarded 'Balaclava' as a battle honour.

The Allies were now able to renew their assault on Sebastopol. Bombardment of the port by land and sea had begun on 8 October 1854 and during the siege the French and British artillery between them hurled 45,000 tons of iron at the fortress. Following the defeat of the Russian attempts to relieve the siege both at Balaclava and again at Inkerman (5 November 1854), the final stage of the siege began. Once again, the 19th Regiment, the Green Howards, fought as part of the Light Division; one of their number, Charles Usherwood, records the order for the assault in his journal, indicating that the desperate nature of the event was anticipated by the issue of extra ammunition:

> The Light & 2nd Divisions will share this important duty and finding respectively the half of each party. The 2nd Brigade Light Division with an equal number of the 2nd Division will form the 1st body of attack, each Division furnishing first a covering party of 100 men under a Field Officer. 2nd a storming party carrying ladders of 160 men under a Field Officer – these men to be selected for this essential duty will be the first to storm after they have placed the ladders ... 10 additional rounds of ammunition will be issued out to each man. The men to parade in Red Coats and forage caps.[12]

The siege of Sebastopol marked a milestone in the use of the rifle as an effective long-range weapon. In a report covering 1855 and 1856, Major Alfred Mordecai of the United States Army comments:

[12] Charles Usherwood's Service Journal

The 19th Regiment (1st Yorkshire North Riding Regiment, the Green Howards), as part of the Light and 2nd Divisions, in the final assault on the Redan at Sebastopol in 1855. In addition to their normal equipment, the men can be seen carrying scaling ladders and fascines (bundles of brushwood) with which to fill ditches. (Painting by Orlan, © Green Howards Museum Trustees)

> ... the protracted siege of Sebastopol served to develop the importance of these arms of long range, as an auxiliary, in both attack and defense of places ... it is only necessary to refer to the extraordinary means used by the besiegers and the besieged to protect their gunners from rifle shots, which could be fired with sufficient precision to enter an embrasure at 500 or 600 yards, and which were effective at much greater distance; or to mention the annoyance and loss caused to the besiegers by Russian riflemen posted in the little advanced entrenchments, commonly called 'rifle pits'.[13]

'Rifle pits' were not a Russian preserve; the British used them as well. Lieutenant-Colonel D. Davidson of the 1st City of Edinburgh Rifle Volunteers, a keen target shooter who later developed a telescopic sight for the Enfield, noted that:

> One soldier was observed lying with his rifle carefully pointed at a distant embrasure, and with his finger on the trigger ready to pull, while by his side lay another with a telescope directed at the same object. He with the telescope was anxiously watching the movement when the [Russian] gunner should show himself, in order that he may give the signal to the other to fire.[14]

[13] Report of a *Military Commission to Europe in 1855 and 1856*
[14] *Army and Navy Journal*, August 1864

Sniping in the Crimea (previous pages)

Men of the Rifle Brigade sniping outside the fortress at Sebastopol. It was probably in the Crimean War that sniping as we understand it today had its beginnings, made possible by the long-range accuracy of the new Enfield rifle in the hands of trained and experienced shooters. Here, in a hastily constructed 'nest' to give concealment and protection, a 'spotter' uses his telescope to observe when an unlucky Russian gunner shows himself through an opening of a gun emplacement in the fortress walls several hundred yards away, then gives the word to the sniper who takes his shot.

The British government later rejected the idea of telescopic sights, but we can see in Davidson's commentary the emergence of the idea of the sniper with his observer, a concept which was developed fully in the century that followed.

The Crimean War is probably the first instance in the history of warfare when the infantry could outgun the artillery. According to Lord Raglan's report of the fighting at Balaclava:

> The 4th Division had advanced close to the heights, and Sir George Cathcart caused one of the redoubts to be reoccupied by the Turks … and he availed himself of the opportunity to assist with his riflemen in silencing two of the enemy's guns. The service was accomplished by Lieut. Godfrey (1st Battalion Rifle Brigade), who proceeding in advance of his battalion with a few men, under cover of a ridge, made such excellent shooting at the Russian gunners (at 600 yards) … that, in Godfrey's own words, 'we got the credit of silencing them'.[15]

This capability of the new rifle was also commented on in a passage in the *Illustrated London News* of December 1854 which tells us that:

> When Prince Menschikoff writes home to the Czar that nothing new has occurred in the siege operations, it must not be inferred that all is going on comfortable in Sebastopol. The latest accounts from the English camp inform us that our riflemen in the trenches continue to pick off the Russian gunners in the most astonishing manner. One rifleman is said to have killed no less than fourteen in one battery, so that we need not be surprised at their bad gunnery. Never was the superiority of skill and science in war so plainly demonstrated as it has been in the use of the Minié rifle, both by the French and English, during the campaign in the Crimea.

Once again we encounter reference to the 'Minié' rifle, adding uncertainty to whether it was indeed the Pattern 1851 Minié or the Pattern 1853 that was actually being used. The same commentator notes with some disdain:

> The Emperor Nicholas, who begins to see that sheer brute force and numbers are hardly a match for such enemies as he has to encounter, has resolved to organise a regiment of sharpshooters for the Spring campaign, 'in full view', he says, 'of the dangers that threaten our beloved fatherland'. If he do[es] not succeed any better in the training of his riflemen than he has done in the manufacture of firearms and powder, there will be not much reason to be afraid of his new regiment.

On 8 September 1855, the siege of Sebastapol came to an end when the Russians were finally defeated and withdrew, blowing up the remainder of the fortifications.

---

[15] Quoted in Busk, H., *The Rifle and How to Use It*

The total number of casualties of the Crimean War is uncertain but is estimated at around 300,000, approximately half of whom died from disease. Today the war is most commonly remembered for the charge of the Light Brigade and their heroic and reckless dash towards the Russian guns, or the work of Florence Nightingale and the pestilence-ridden hospital at Scutari, or the instigation of the Victoria Cross as the highest British military award for valour in the face of the enemy. But perhaps the most important aspect was the fact that for the first time in the history of warfare, the gunners of the artillery were at the mercy of the infantry who, with their new, accurate, long-range rifles, could pick them off with impunity while remaining out of range of the guns.

THE INDIAN MUTINY OR 'GREAT REBELLION'

By 1857, the East India Company, which had its origins in a Charter signed by Queen Elizabeth I in 1600, basically governed India on behalf of Britain. It had its own army, along with a few regular British Army units, controlled by the Governor-General. The three Company armies – Bengal, Madras and Bombay – totalled around 233,000 Indian troops and 36,000 British, all equipped identically, and commanded by a British officer corps.

There had been growing unrest but this was turned into violent protest by the introduction of the Enfield rifle – or, more accurately, not by the rifle itself but by the cartridges used with it. We are told by Arthur Moffatt Lang, a young officer of the Bengal Engineers, in his journal that:

> ... a mutinous spirit has been pervading the Bengal Regiments: that at Dinajpur, Lucknow, Meerut, Phillaur and Ambala, regiments have been either mutinying or burning. Well today I have just heard from Ambala that two Native Infantry regiments mutinied to save some of their comrades imprisoned for not using the greased cartridges.[16]

That the uprising should have begun in this region of India is undoubtedly because it was the Bengal regiments that were first issued with the new rifle and cartridge. These paper cartridges contained a bullet lubricated with a grease which was very probably a combination of beef and pork tallow. The way the rifle was loaded required the cartridge to be clamped between the teeth, the paper portion containing the powder and bullet torn away, and the powder poured down the barrel followed by the bullet. Since the cow was sacred to the Hindus and pork was considered unclean by the Muslims, the very fact that they had to handle animal grease and risk putting it into the mouth seriously offended both religious groups.

East India Company officers had become aware of possible trouble over the cartridges in January when, apparently, a labourer at Dum-Dum arsenal in Bengal near Calcutta had taunted a high-caste sepoy by saying he had lost caste by having bitten a cartridge. In fact, the arsenal at Dum-Dum had not

[16] Lang, Arthur Moffatt, *Lahore to Lucknow: The Indian Mutiny Journal of Arthur Moffatt Lang*

actually begun production of the new cartridge and not even a practice shot had been fired. Nevertheless, it appears that on 27 January, Colonel Richard Birch, the Military Secretary, ordered that all cartridges issued from depots were to be grease-free and that sepoys could grease them themselves using whatever mixture 'they may prefer'. This seems only to have convinced many sepoys that the rumours were true and that their earlier fears were justified!

Another step towards conflict occurred on 29 March 1857 at the Barrackpore (now Barrackpur) parade ground, near Calcutta. Mangal Pandey, a 29-year-old member of the 34th Bengal Native Infantry stationed there, had been angered by the East India Company's interference in long-established Indian tradition by, to all intents and purposes, stealing the lands once belonging to the Nawab of Bengal, Sayyud Mansa Ali, and removing the Nawab's authority. For those, like Pandey, who had once been subjects of the Nawab, this meant the loss of their right to direct appeal to the British Resident at Lucknow for justice – a significant privilege in the context of native courts – and this was felt to be an affront to their honour. Pandey declared to his fellows that he would fire on the first Englishman he saw. This turned out to be the adjutant, Lieutenant Baugh, who came out to investigate the unrest. Pandey did open fire but hit the horse instead. In the affray which followed, Pandey eventually turned his rifle on himself. He did not die, but was later brought to trial and sentenced to death. Pandey is revered by many Indians as a martyr and is often cited as the person who really initiated the uprising – indeed, Lang in his journal frequently refers to the rebels as 'Pandies'.

Things escalated a few weeks later in Meerut, when on 24 April, 85 Indian cavalry troopers at the garrison refused to use the new cartridge and were disgraced and imprisoned. While the British units were at church, the comrades of the prisoners released them and went on the rampage, killing as many British and European men, women and children as they could find.

Military buildings, encampments and personnel were among the main targets of the mutineers at Meerut but women, children and civilian buildings were not exempt. (*Illustrated London News*, July 1857/Private collection)

Before the British contingent could respond, the rebels had fled to Delhi. The following day, the butchery continued there and a few British officers and men held the arsenal as long as possible before taking the suicidal step of blowing up both the ammunition magazine and themselves with it, in the process killing and wounding many of the attackers.

Literature often refers to the massacre at Cawnpore. In fact there were two, and possibly even three massacres. The first occurred when Dandu Panth (Nana Sahib), Rajah of Bitpur, led the rebellious native troops in the three-week siege of the British contingent in June 1857. He eventually persuaded Sir Hugh Wheeler, the British commander, to surrender his handful of troops and nearly 200 non-combatants (largely women and children) with the promise of safe conduct to Allahabad. While the men were embarking on river boats, they were murdered and the women and children imprisoned.

In July, General Sir Henry Havelock left Allahabad with 2,500 men for the relief of Lucknow but he had to relieve Cawnpore first. After a decisive battle, the bodies of the murdered women and children were discovered thrown into a well. Nana Sahib managed to escape the wrath of the British soldiers but most of his followers were not so fortunate.

In rejecting the new cartridge and the weapon for which it was designed, the rebels denied themselves a superior weapon to the ones they were equipped with, the old smoothbore muskets. Despite their superiority in numbers and their early success, they had placed themselves at an immediate disadvantage which was ultimately to lead to their downfall. Lang spent some time at the old fort at Jalalabad preparing siege materials and his account of an attack there by some rebels illustrates the power of the Enfield rifle:

The brutality and duplicity exhibited by the rebels at Cawnpore, shown here killing unarmed civilians, probably did much to inspire revulsion among even some sympathizers and, within the British military especially, a desire for savage revenge. (Private collection)

The charge of the Highlanders at Cawnpore, led by General Havelock. After marching 126 miles in nine days during the hottest season, immediately followed by two battles at Fatehpur and Aong, Havelock's forces finally shattered Nana Sahib's rebel army at Cawnpore. (Private collection)

> ... masses of infantry with scaling ladders came on ... but they could not stand the penetration of our bullets, which would kill two men one behind the other. Those 'Minies' and 'Lancasters' do give us a wonderful advantage over Pandy, and he finds it very disagreeable to find the conical bullets dropping in at 1500 yards into his columns. He can hardly ever get the pluck up to venture within range of his own muskets ... must now mourn the folly of rejecting these same cartridges which play such mischief with him.[17]

If the massacre at Cawnpore is branded into history, then so too is the relief of Lucknow. Fierce fighting took place at the Sikander Bargh (also known as Sikandar Bagh, Sikandra Bagh or Secundra Bagh), a walled garden on the outskirts of Lucknow built for the Nawab of Awadh as a summer house and taking its name from the Nawab's favourite wife, Sikander Mahal Begum. Once again we can call upon Lang for an account, in a passage which vividly describes how all types of weapon, from artillery to 'axes and muskets' were employed:

> ... then began our work, some of our Sappers knocking down a mud wall so that a big 24 pounder might play from the street at the Sikandar Bagh, an enclosed garden held by some 2,000 Pandies. Then we were at work in making ramps from the street up some banks, up which we pulled more heavy guns ... in a little time we had several heavy guns and some light guns smashing away at the Sikandar Bagh, infantry lying under shelter ready for a rush ... Pandy all the while keeping up no end of a fire ... I saw Lt Paul rush ahead waving his sword, and the 4th Punjabis, yelling and shouting, as they charged along behind him ... the effect was electrical ... we rushed along too; up sprung the 93rd and 53rd and, cheering and shouting 'remember Cawnpore' ... axes and muskets soon smashed in the gate, and then didn't we get revenge ... the first good revenge I have seen.[18]

[17] Lang, Arthur Moffatt, *Lahore to Lucknow: The Indian Mutiny Journal of Arthur Moffatt Lang*
[18] Ibid.

The desperate storming of Sikander Bargh was led by the same 93rd Regiment, armed with their Enfield rifles and bayonets, which had performed such valiant deeds at Balaclava. (Courtesy of the Council of the National Army Museum, London)

That passage serves to show how visceral war can be, even for the officers. And it does not end there:

> ... such a distracting row of thousands of rifles being fired without intermission I never heard, and such a sight of slaughter I never saw ... Pandies were shot down and bayoneted in heaps three or four feet deep ... at the house in the middle of the rear wall and in the semicircular court beyond ... they shut the many thin doors and thousands of bullets were poured in ... It was a glorious sight to see the mass of bodies, dead and wounded, when we did get in ... the mass of bodies were set fire to, and to hear the living as they caught fire calling out in agony to be shot was horrible even in that scene where all men's worst passions were excited. The bodies have now been counted and buried – 1,840 Pandies.[19]

Clearly the Enfields' bayonets as well as their bullets were in use in this engagement.

The violent uprising continued for just over a year and was finally brought to more or less of a close with the battle at Gwalior in June 1858. Retribution against the surviving rebels was varied and though many escaped the consequences, others were brought to trial in military courts martial. Some of those found guilty faced the death penalty by firing squad or hanging. Others faced a far more ghastly execution – being 'blown from the guns' – which entailed standing the rebel in front of the muzzle of a field gun, to which

[19] Ibid.

Convicted rebels being 'blown from the guns'. The consequences are too ghastly to contemplate and make No. 1 Field Punishment (being strapped spread-eagled to a carriage wheel and flogged), and even a firing squad, seem mild by comparison. (Private collection)

he was tied, and then firing it. It was, without doubt, the most wilfully brutal and barbaric punishment meted out to some of the convicted rebels.

That the cartridge caused the Mutiny, or Great Rebellion, is claiming too much. That it was the last straw in a series of perceived injustices is more likely. Issuing Indian troops with a cartridge of that nature might even be thought to have shown a lack of consideration for religious sensitivities. But when the specifications for the cartridge were being drawn up at Woolwich, the possible consequences, several thousand miles away, of using tallow were probably never even thought of. Tallow was simply a useful, cheap and readily available lubricant. S. S. Thorburn was one of those very few Britons who admitted the superiority of the Enfield rifle and the decisive part it played in the sepoys' defeat in 1857. In the Appendix of his book he makes a very telling observation, echoing what Lang had already said: 'Had the sepoys accepted the Enfield and mutinied afterwards, our difficulties in suppressing their revolt would have been enormously increased.'[20]

In 1859 tallow was abandoned as a lubricant and was replaced by beeswax, not on religious grounds necessarily but because tallow was found to cause some corrosion to the bullet. At least, that was the official reason!

THE AMERICAN CIVIL WAR

The next major conflict in which the Pattern 1853 saw widespread use was the American Civil War, in which a nation was divided over the question of slavery. When Abraham Lincoln won the Presidency in 1860 with the backing of all the Northern, and numerically larger, 'anti-slavery' states, the writing on the wall was clear to the seven lower Southern 'slavery states' and during the winter of 1860–61 they seceded from the Union, forming their own government with its own constitution under Jefferson Davis. When Lincoln came to take his Presidential oath to 'preserve, protect and defend the United States and its Constitution', the states were no longer united and civil war was inevitable.

[20] Thorburn, S. S., *The Punjab in Peace and War*

Some have suggested that the Civil War was the first in which rifles were used extensively. That this is not quite true should already be apparent, the Crimean War having that distinction. However, the Civil War was on a far greater scale than either the Crimean War or the Indian Mutiny, or even both put together. In fact, its scale was staggering. It was fought, theoretically, over an area approaching the size of India and Europe put together. Around 600,000 Americans were to die in it, more than in *all* the country's other wars combined. In simple numerical terms it is safe to say that more rifles were indeed used in the Civil War and though it may not have been the first major war in which rifles were extensively used, it was certainly the largest and was probably the last in which muzzle-loading firearms of any kind were the main infantry weapon.

Like Britain at the start of the Crimean War, America at the onset of the Civil War was unprepared. Before the war, the Springfield Armory in the North was producing only 10,000 rifles per year; by 1864 this had risen dramatically to 300,000. The sudden process of enlistment created the same problem for both sides – more soldiers than weapons – with the result that many did not even have firearms to train with and many who did often had to make do with old flintlock muskets converted to percussion. The 2nd Mississippi Regiment received some 'old army muskets' which failed to inspire confidence – 'why such a weapon was ever dealt us with which to fight the enemy is a puzzle to me, as there is about equal danger at either end.'

The state of Massachusetts, home to the Springfield Armory, was very aware of this problem and sent emissaries to England and Europe to buy Enfields. The records of the 10th Massachusetts Volunteers of Springfield show that in July 1861:

> … Friday morning the regiment marched to the US Armory and returned the muskets loaned them for the purpose of drill, and in the afternoon we received our full supply of the Enfield rifled musket. For this the Regiment may well thank our efficient Colonel, whose influence has procured for us so fine an arm; whilst other Regiments are obliged to take the guns we returned, (smooth bore muskets of the old model.) The Enfield gun, purchased by the State in England, though differing in many respects from the Springfield rifled musket, is a handsome and no doubt serviceable weapon, and I think fully equal to the Springfield arm. It is browned, so that no burnishing is required to keep it from rusting, and a more correct aim can be obtained in a bright sun than with a polished barrel.[21]

In the case of the 20th Massachusetts:

> … our old smoothbore muskets were exchanged for Enfield rifles, which were carried for the remainder of the war… It was not considered quite so good a gun as the new Springfield rifled musket, but it was a good rifle and decidedly superior to any of the other guns given to our soldiers.

[21] Both quotes from Hess, Earl J., *The Rifle Musket in Civil War Combat*

> Many, even of the Massachusetts troops, had only old smoothbore Springfield muskets, recently altered from flint-lock to percussion, owing to the impossibility of getting a sufficient number of rifles of any kind.[22]

Some regiments even continued using smoothbore muskets throughout the war. There were also many who took their own guns with them – such as shotguns and squirrel rifles – and both sides appealed to the populace for weapons. Not only were shotguns and squirrel rifles ill-suited to the battlefield, they simply added to the variety of weapons being used with the result that many of the early engagements at least were undertaken with a motley collection of arms that increased the problem of ammunition supply.

It is perhaps ironic that the Enfield rifle should find far greater usage in the American Civil War than in any wars engaged in by Britain. Around one million of them were exported in roughly equal measure to both sides by the private arms trade in London and Birmingham and, undoubtedly, from Liège also. In many cases they perhaps did not fully conform to British military specification and quite probably many were assembled from parts that, though serviceable, came from rifles which had been sold out of store by the government as damaged, worn or obsolete, or from serviceable parts which had been rejected in the rigorous but sometimes unfathomable government inspection process. Who better to refurbish them for resale than those who created them in the first place? Importantly, they were supplied by the private trade, not by the British government. This point was made very clear at the time in a note in the *Scientific American*:

> We have also seen it stated in several papers that Enfield Rifles, purchased in England, have arrived here for arming our volunteers. This is also a mistake. The rifles made at Enfield are all for the British army, because the works belong to the government. Such rifles cannot be sold to private parties, nor obtained upon any account from England. The British rifle muskets that have been imported are equally as good as those made at Enfield, because they are similar in pattern, though they are manufactured by private gunsmiths.

This raises an interesting point, especially from the collector's point of view. The rifles supplied during the Civil War can easily carry a number of apparently conflicting markings – most barrels were re-proofed and so could have Birmingham or London proof marks, possibly in addition to original military proof marks; they often have inspector's marks from a previous existence when they *were* part of the British military arsenal; they may have unmarked items manufactured new by the trade at the time for sale to America, they may have original old stocks refurbished by contractors, or in their original state if good enough, with British military markings, or newly made with no markings, or with private manufacturer's or even agent's markings – the scope for variation is almost endless.

---

[22] Bruce, George A., Brev Lt-Col, *The Twentieth Regiment of Massachusetts Volunteer Infantry, 1861–1865*

The storming of Fort Wagner by the 54th Massachusetts Volunteer Infantry Regiment is a forcible reminder that the Civil War was a war for the freedom of African-Americans. (Private collection)

In that sense, therefore, such rifles are not *true* Enfield Pattern 1853s, even though they may look identical, because they were no longer part of the British military service with its rigorous inspection routine and were not supplied directly from the British military establishment.

How they were distributed once they reached America is uncertain. One regiment that definitely used them and achieved fame during the Civil War was the 54th Massachusetts Volunteer Infantry, the first all African-American regiment to be raised in the Union. Though the desirability or efficacy of using African-Americans as combat troops was questioned by some, once the regiment was in front-line service it quickly distinguished itself. Its most famous exploit was the storming of Fort Wagner on 18 July 1863, in which action its colonel, Robert Gould Shaw, was killed and the young sergeant William Carney, though badly wounded, saved the colours. For this act he became the first African-American to be awarded the Congressional Medal of Honor. The story of the 54th and this action is captured in the film *Glory*.

But half a million rifles go a long way in equipping an army and many regiments, both Union and Confederate, used them in many engagements. However, the North had greater access to American-made arms, whereas the South, which had very limited arms manufacturing facilities, probably used some Enfields in every engagement. One wealthy Southern planter joined the 10th South Carolina as a private and used his own money to buy his

Storming Fort Wagner (previous pages)

To reach the ramparts of Fort Wagner, the 54th advanced first through a barrage of shot, shell and canister, and then a terrible onslaught of close-range musketry, all the while having to avoid the natural or man-made obstacles which impeded their progress and made them easier targets. By the time they reached the ramparts where hand-to-hand fighting was necessary, their numbers had been so reduced that they were soon forced to retire. Starting out 650-strong, the assault on Fort Wagner cost them almost half their men. The assault failed, but the regiment and all it stood for won its place in history.

comrades 155 Enfield rifles. Not surprisingly, he was later elected Captain of the Company! Indeed, in 1862 General Gorgas, Chief of Ordnance of the Confederate States, instructed that the standard rifle cartridge of the Confederacy was to be that of the Enfield rifle and what manufacturing facilities existed were to duplicate the Enfield rifle. Nevertheless, it seems that the Confederacy also adopted the US Model 1855 rifled musket, or at least a version of it. In James Burton they had the very man to implement and oversee the manufacturing process, whichever rifle was produced. Not only did he have drawings of the Enfield rifle and the necessary machines to manufacture it, he was probably the only man in America who had actually done such a thing, having overseen the setting up and implementation of manufacture at the Royal Small Arms Manufactory at Enfield! A huge armoury was built at Macon, Georgia, for the purpose and much of the machinery for making both arms and ammunition was purchased by Burton from Greenwood and Batley in Leeds. Burton already knew them since it was the company which had copied some of the Robbins and Lawrence machines to finish equipping the factory at Enfield. It is interesting that Thomas Greenwood was an ardent supporter of the Confederacy.

Detail of St Gaudens' monument to the 54th Massachusetts, clearly showing the young volunteers with their Enfield rifles. (Author's collection)

Unlike the Crimea or the Indian Mutiny, where the British troops obviously used British-made weapons, the mixture and distribution of arms encountered in the Civil War is not recorded in the official histories. Reliance has to be placed on the few surviving first-hand accounts in the form of soldiers' letters, or on material recovered from battlefield sites, to obtain definitive knowledge of which weapons were being used. This does not always tell us which side was using them since we know that both sides regularly scavenged the battlefields for useful equipment after an engagement.

One of the early major battles was that at Shiloh on 6 and 7 April 1862. It was also one of the bloodiest. General Grant and President Lincoln had decide to deal a decisive blow in the Confederate heartlands, but the Confederate forces made a surprise attack on Grant's Union encampment which was poised on the banks of the Tennessee River. Almost 100,000 men engaged in the battle that raged around the small Shiloh Church, which ironically gains its name from a Hebrew word meaning 'place of peace'. One of these battles took place in Sarah Bell's peach orchard, a wooded thicket in which Union troops came under such tremendous fire from Confederate forces that it was called the 'Hornet's Nest'.

Henry Morton Stanley, who later gained fame from his search for Dr David Livingstone, was in the Confederate army at this battle and recalls:

> We loaded our muskets, and arranged our cartridge pouches ready for use. Our weapons were the obsolete flintlocks and the ammunition was rolled in cartridge-paper, which contained powder, a round ball, and three buckshot. When we loaded we had to tear the paper with our teeth, empty a little powder into the pan, lock it, empty the rest of the powder into the barrel, press paper and ball into the muzzle, and ram home.

The 'Hornet's Nest' at the battle of Shiloh, so named because of the intensity of fire from Confederate Enfields and the almost continuous sound of bullets buzzing through the air. (Private collection)

From this it could be inferred that the South was very poorly equipped, having to use such antiquated arms. However, one Union soldier, Henry Dwight of the 20th Ohio, observes that 'his regiment exchanges their altered muskets [smoothbore flintlocks converted to percussion] for Rebel Enfields marked "V. R." with the British crown on the lock following Shiloh' and that 'just after Shiloh, Federal Colonel Morgan L. Smith reported capturing 14 entirely new [Enfield] rifles, with the Tower stamp, 1861'. In another instance in early June 1862, Dan Sickles, commanding the Federal Excelsior Brigade, boasted that after the battle of Fair Oaks 'the fields were strewn with [Confederate] Enfield rifles marked "Tower 1862", and muskets marked Virginia'.[23] All these commentaries show that some Confederates at least had high-quality rifles and were often better equipped than some of the Union forces.

After the two-day battle almost 25,000 men were dead, wounded or missing; although General Grant had been able, after receiving reinforcements, to snatch a victory from the jaws of defeat, it had been an expensive one and served to show that the war was not going to be won either quickly or cheaply.

The engagement at Antietam in September was to prove even more costly. In fact, 17 September 1862 has been characterized as the bloodiest single day in the whole war, one which left nearly 23,000 dead and wounded, more than four times the number of casualties on the Americans' Normandy beaches of D-Day, 6 June 1944. A short while later, at Fredericksburg, the rallying cry of officers of the 69th New York Volunteer Infantry – the 'Irish Brigade' – was 'Come on boys, this is nothing to Antietam'.

Some units, such as the 7th West Virginia, were armed with the British-made Enfield rifled musket, but the New York regiments of the Irish Brigade were issued the Model 1842 .69cal smoothbore musket. This was actually a weapon favoured by the commander of the brigade, since it

[23] Quoted in Castel, Albert, ed., 'The War Album of Henry Dwight'

A group of Civil War, possibly Confederate, soldiers at an unknown location, proudly posing with their Enfield rifles. (Library of Congress)

could fire 'buck and ball' (a load of buckshot and musket ball) at close range with deadly effect. Even so, one officer of the Irish Brigade, Captain Felix Duffy, who was always ready to pick a fight, 'liked to carry an Enfield rifle musket into battle in order to have a crack at the enemy'.

In the early morning near Dunker Church, at the edge of David Miller's cornfield, Union troops began to exchange volleys with the Confederates. Eventually, 'the Federals locked bayonets onto their Enfield rifles and charged'. The Confederate regiment, 250 men strong when the morning began, brought a mere 24 soldiers out of the cornfield. The Union did not escape lightly; it suffered 50 per cent casualties and lost all of its field officers. Clearly, for all its improvements over its predecessors in long-range accuracy, the Enfield still proved useful in a traditional bayonet charge.

Around midday, a mile or so to the south, Union forces attacked and overran the Confederate central line of defence which was a sunken road. After repeated Union attacks, the defenders abandoned the sunken road, but so appalling had been their casualties that it was christened the 'Bloody Lane' – a ditch strewn with the bodies of the dead and dying. General Hooker later commented that 'it was never my fortune to witness a more bloody, dismal battlefield'. One Confederate soldier, Hill Fitzpatrick, wrote to his wife: 'To my greatest mortification my gun [most likely an Enfield] failed to fire, but I soon recollected that it was charged with a Yankee cartridge which had to be ram[m]ed hard. I drew my stick gave it a hard ram tried again and went clear as a whistle.'[24] It is worth recalling that the 'Yankee' cartridge was different from the standard Enfield cartridge (which Gorgas had declared to be the standard Confederate cartridge) in that it was slightly larger. To counteract such errors General Lee issued instructions that 'The .58 caliber will be inserted naked', that is without the paper wrapping, as distinct from the Enfield cartridge where the wrapping was retained when loading.

[24] Glatthaar, Joseph T., *General Lee's Army: From Victory to Collapse*

Many at the time acclaimed the Enfield to be one of the best rifles available in the Civil War, and this is supported by comments from both Union and Confederate soldiers. In his diary, Confederate W. H. Tunnard records:

> 28th. (May 1863) Still clear and warm. A courier succeeded in reaching the city (Vicksburg, Mississippi) with 18,000 caps which were much needed. Heretofore, the Third Louisiana were armed with the Confederate Mississippi rifles furnished them at Snyder's Mills. These arms were almost worthless, often exploding, and so inefficient that the enemy boldly exposed themselves, and taunted the men for their unskilful shooting. On this day, however, the regiment was supplied with Enfield rifles, English Manufacture, and Ely's cartridges, containing a peculiarly shaped elongated ball, of the finest English rifle powder.[25]

Sharpshooters on both sides approved of the Enfield for its range and accuracy though it was probably more favoured by the South since they had little of matching quality. One Confederate commentator noted that:

> Every short [two-band] Enfield which came into possession of any of our men was taken away and given to these men [sharpshooters] … But there were not enough and some of them had the common long Enfield. Both kinds had a long range and were effective. The short guns were given them as they were lighter and handier.[26]

At the battle of Thorough Gap in August 1862, the Enfield proved its worth against artillery, when Colonel Henry L. Benning deployed all of his men of the 2nd and 17th Georgia armed with 'long range guns' (Enfield rifles) to fire on a Union battery posted 400 to 500 yards distant. 'Under our stinging fire, the battery was driven off despite the commanding position of those field guns.'[27]

Some in the North also found the Enfield a weapon of merit:

> In early 1863 rear Admiral David Porter requested from Grant 800 soldiers for the vessels. By March, the army had supplied the 58th Ohio Infantry for the USS ships *Mound City*, *Signal*, *Carondelet*, *Baron deKalb*, *Benton*, *Pittsburg*, *Linden* and the *Louisville*. Companies from the 101st Illinois, 39th Illinois and the 20th Illinois Infantry were also assigned to Porter's vessels. These infantrymen, armed with their Enfield rifles, were used in place of Marines for sharpshooters and to help man the guns.[28]

The Enfield did, however, have its detractors, especially in the North where national pride would naturally favour the Springfield. This is supported by at least one commentator in the 37th Massachusetts Regiment who notes:

[25] Tunnard, W. H., *A Southern Record*
[26] Bradwell, G., *Under the Southern Cross*
[27] Quoted in Sword, Wiley, *Firepower From Abroad*
[28] McAulay, John D., *Civil War Small Arms of the U.S. Navy and Marine Corps*

The Springfield muskets with which the regiment was equipped were issued ... [and] were received by the men with delight. Not only were they without doubt the most efficient muzzle-loading military rifles ever made ... there was a patriotic aversion to the foreign-made weapons with which some of our troops were armed.[29]

In the case of the 34th Massachusetts Infantry we are told, 'we have had in camp for some time, Enfield Rifles, which have been sent up for our use; but the Lieutenant Colonel Commanding keeps them boxed up, refusing to issue them until it is established certainly that Springfield rifles cannot be procured'.[30]

In some rare cases this aversion turned to destructiveness, as with the 24th Massachusetts: 'When, fixed with bayonet, the latter was driven into the ground, and the gun pulled over at right angles, all concerned deemed the weapon defective, and were better satisfied when rifles of Springfield make were placed in their hands'.[31] But, before any conclusions regarding faulty workmanship are drawn, we are left with two vital questions unanswered – where did the bend occur and how much force was required to cause the bending? In other words, was it a failing likely to have occurred in normal combat use? We shall never know the answer to that.

The 10th Massachusetts Infantry:

Received supply of Enfield rifles and returned to the armory the old pattern muskets [1842 Model] which had been used for drill. The Enfield received would not compare favourably with the Springfield musket, new pattern. The workmanship was rough and they were poorly rifled, and the parts would not interchange like the American gun. It was necessary to keep an armourer with the Regiment, to fit such parts of the musket as were accidentally broken in service.[32]

This comment is particularly interesting in highlighting the lack of interchangeability of the Enfields, because in 1857 Enfield began producing the version of the Pattern 1853 which was interchangeable and gradually the non-interchangeable weapons became obsolete and were sold off to the private gun trade.

The history of the Civil War resounds with the names of many people, places and battles that have become famous worldwide, but one common denominator is that both sides used the British Pattern 1853 Enfield rifle extensively in its various forms. Another is, of course, that like any war, it left vast swathes of death in its wake, destroying not only individuals but families and communities also. In retrospect it perhaps gives a vague hint as to what was to happen in Europe half a century later.

---

[29] Quoted in Bowen, James L., *History of the Thirty-Seventh Regiment, Mass. Volunteers*
[30] Lincoln, William S., *Life with the Thirty Fourth Mass. Infantry in the War of the Rebellion*
[31] Roe, Alfred S., *The Twenty Fourth Regiment, Massachusetts Volunteers, 1861–1866*
[32] Newell, Joseph K., Capt, *Annals of 10th Regiment*

IMPACT

Accuracy, industry and surgery

The Pattern 1853 rifle could not fail to have major impacts, and not all of these were on the battlefield. However, that of course is what really mattered, and where the Enfield fully proved itself.

THE SCHOOL OF MUSKETRY

In one sense the Pattern 1853 rifle brought a level of equality into the army since every soldier – not just an elite and selected few in the various rifle regiments – now had a high-quality, long-range and accurate weapon. But because of this it created another demand in its wake. The 'musket' had been around in the army for so long that it had become something that was taken for granted. No great skill was needed in its use, beyond taking care on safety grounds and keeping it clean. Loading was simple and straightforward; firing was a matter of holding one's nerve until the enemy was close enough and the command to fire was given. The rifle presented a whole new set of demands and capabilities, and to ensure its potential for long-range accuracy was fully realized, in June 1853 Lord Hardinge established the School of Musketry in what had been the barracks of the Royal Staff Corps at Hythe in Kent; however, it did not come into operation until the spring of 1854.

The school was not intended to train every soldier in the army, only selected officers and NCOs from the regiments, who would then serve as regimental instructors. The course had two elements: drill and practice. Drill encompassed the theory of musketry, covering aiming, correct positioning of the body, judging distances, arms drills, cleaning arms and the making of cartridges. After the soldiers had been thoroughly drilled,

Aiming drill; instructors being trained in the correct aiming of the rifle at ranges from 100 to 900 yards. (*Illustrated London News*, October 1855/Private collection)

this training was put to the test by the practical application of the theory, especially target practice and judging distances. The target practice included firing individually at ranges from 100 yards to 900 yards in 50-yard increments, as per the drill instruction, in small groups of men, formed according to merit; platoon firing in file and volleys; and firing in extended order as skirmishers, in which instances the art of judging distances and making appropriate sight adjustments was put into practice. All of these were new skills to officers and men accustomed to the use of the old smoothbore muskets. The loading drill was very similar to what had been carried out in the past, but the major difference was that the bullet had to be loaded in the correct orientation; i.e., point forward. An almost identical training system was adopted by the United States following the introduction of the Springfield rifle.

Those selected to attend the School of Musketry, one must suppose, would be the best able to benefit and to pass on that training to the men of their regiment. But training in the correct loading and handling of the rifle is one thing. Training in marksmanship is another and, unless very apt pupils existed, it needed to be intensive and prolonged in order to reach

Position drill; correct positioning of the body when firing was an important part of achieving good marksmanship. (*Illustrated London News*, October 1855/Private collection)

Judging distances; to be able to judge distances accurately was vital if effective shooting was to be achieved. (*Illustrated London News*, October 1855/Private collection)

any level of excellence. This could hardly be achieved with an allocation of 90 rounds per man per year, even on top of the 110 rounds allowed for new recruits, and that practice was only allowed on stations where a range of at least 300 yards existed. Even today, where these rifles are still used for target practice, shooters can easily expend a 'year's' supply in a few days and expect to expend much more each year in the pursuit of perfection. Such a skimpy approach to marksmanship training by the authorities, even if simply on financial grounds, perhaps underlines the lack of a full appreciation of the capabilities of the new rifle and how it might be most effectively used in warfare.

THE VOLUNTEER MOVEMENT AND THE NRA

Despite the alliances established during the Crimean War, shortly after its conclusion the Chancellor of the Exchequer noted that 'when the present government assumed office, our pacific relations with France were not a matter of weeks or days, but of hours'. A correspondent for the *Military Spectator* wrote from Paris in August 1858 that the French army numbered upwards of 430,000 men who were 'well armed, well clothed, and the whole of their appointments are of the best description'.

Alongside the confusion about what was actually happening in, and intended by, France, there seemed also to be a threat that the United States might take up arms against Britain. In the midst of this, 25,000 troops were being sent to India in the aftermath of the Mutiny, leaving the country almost devoid of protection. And yet, in the face of such possible difficulties, we are told that the government was rapidly disbanding the militia which remained virtually the only means of home defence.

It was these factors, plus the very important consideration that a new, high-quality and very efficient rifle had been introduced into the service, that stimulated the development of the Volunteer movement. There was a very distinct difference between the Militia and the Volunteers. The rank-and-file of the former tended to be drawn from the poorer classes of society; the Volunteer movement, on the other hand, would be based on the premise that 'the men composing it should, without exception, be those who by birth or position have a stake in the country; whose loyalty is beyond question; who would have the strongest possible interest not only to repel aggression, but to maintain tranquillity, and to defend and uphold all established institutions'.[33]

Such men would be 'for the most part by birth and education gentlemen'. And of course, the feature that would attract such men would be the rifle, which in its handling and accurate shooting required at the least good judgement, but also – in the acquisition of good judgement – the capacity to grasp the basics of ballistics. In other words, it required men of a high degree of education. Not only that, but the Volunteer movement brought the opportunity to create a new shooting sport.

The nature of the status of such regiments is indicated by the Victoria Rifles, formed around 1854. Its colonel was none other than the Duke of Wellington and among the members enlisted in one year alone were four Members of Parliament, 30 graduates of Oxford and Cambridge universities, and 28 Members of the Bar plus army officers and a county magistrate or two.

That the Volunteer movement flourished is a matter of history. But it was greatly aided by the formation, on 16 November 1859, of the National Rifle Association – not to be confused with its American counterpart which was formed 12 years later. In the words of Lord Elcho in a letter to *The Times* on 9 December 1859:

> The National Association is formed 'for the encouragement of Volunteer Rifle Corps, and the promotion of Rifle-shooting throughout Great Britain' by raising funds for the establishment of a great annual national meeting for Rifle-shooting, similar to the 'Tir Fédéral', which takes place every two years in Switzerland, at which prizes will be competed for. The principal prizes will be opened only to enrolled effective volunteers, and it is thus that encouragement will be given to the volunteer movement: but, at the same time, with a view to promote Rifle-shooting as a national pastime and custom, it is proposed likewise to establish prizes which will be open to all comers, whether volunteers or not.

Queen Victoria opened the first meeting of the National Rifle Association at Wimbledon in 1860 by firing a Whitworth rifle from a mechanical rest, scoring a bull's eye at a range of 400 yards.

The competition meetings proposed by Lord Elcho and opened by Queen Victoria continue to this day, known as the 'Imperial Meeting'. In 1890 the Association received its Royal Charter and its list of vice presidents alone reads like an entry in *Burke's Peerage*. And, while a variety of muzzle-loading rifles were used from the outset, all of this was made possible only by the introduction of the Enfield rifle into general military service.

---

[33] Busk, H., *The Rifle and How to Use It*

Scenes such as this ensured that no Native Infantry were issued the rifled version of the Pattern 1853 in the aftermath of the Indian Mutiny. (Private collection)

THE INDIAN MUTINY

It has already been suggested that the Pattern 1853 rifle cannot really be considered the sole cause of the Mutiny, although its associated ammunition was, legitimately or not, a contributory factor. However, the Indian troops' rejection of the new cartridge meant they were significantly less well armed and so the British victory can be attributed, in part, to the new rifle. Having said that, one of the outcomes of the Mutiny itself was, as already discussed, a new family of weapons based upon the Pattern 1853 rifle but with the important difference that they were smoothbored, maintaining the superiority of the weaponry in the hands of British troops.

ARTILLERY

Artillery at the time of the Crimean War was considered to be at the zenith of perfection, yet it was suddenly shown to be outgunned by infantry armed with rifled muskets and to be relatively ineffective against the fortress of Sebastopol. To increase range for greater effect against infantry at a greater distance, or to increase the effect of projectiles against substantial structures, greater muzzle energy was needed. This in itself meant the use of larger quantities of powder but that could simply not be accomplished with even the best guns available. Increasing the powder charge basically meant increasing the pressure generated inside the bore of the gun, and that they could not withstand. They would simply burst. What was required was a completely new form of artillery and this brought about a reappraisal of artillery design and construction, based on a fuller understanding of the properties of the materials used and how they behaved under such conditions.

This process was initiated by none other than a civil engineer, Robert Mallet. It was to transform gunmaking from an 'art' into a science. Rather than adopting the traditional artillerist's approach, he 'reverse engineered' by looking at – particularly in regard to Sebastopol – the effect required on the structure to be bombarded, the nature of that structure, and the nature and size of the projectile required to achieve the required effect. He then set about designing a gun capable of delivering it. To do this, applying mathematics along with a sound knowledge of materials, he developed a completely novel gun design and method of construction for the building of a huge mortar with which to bombard Sebastopol. The gun was not built in time to see service in the war but it was tested and Mallet's design principles and mathematical reasonings were verified. Mallet's mortar, with its 36in bore, remains the largest-calibre 'conventional' artillery piece ever to be built and, although never fired in anger, it provided the foundations for all artillery that was to follow. The days of the cast iron or bronze guns were over and a revolution was about to begin. And one is tempted to wonder what influence proximity to this work may have had on Robert Mallet's son, John, who was later to become an important figure in the Ordnance Bureau of the Confederate States of America during the Civil War.

MANUFACTURE

Perhaps one of the more important, but often overlooked, impacts of the Enfield rifle was not in the way it was used but in the way it was made. When it first went into production, the traditional course was followed and in October 1853 orders began to be placed with private contractors in Birmingham for the variety of component parts which would then be shipped into store at the Tower of London. From here they would be issued to other contractors for assembly, or 'setting up'.

But two exhibitors at the Great 'Crystal Palace' Exhibition in London in 1851 had shown that there were other ways of doing things. One company was owned by that flamboyant salesman, showman, inventor and entrepreneur Samuel Colt, whose stand was festooned with revolvers of all kinds, all made by machinery. So pleased was Colt with his success at the exhibition that he soon established his own factory in London where guns being made by machines instead of craftsmen could be seen for the first time. The other was a company little known outside America and probably little known outside the small town of Windsor in Vermont where it was based. The company was Robbins and Lawrence and they simply exhibited six rifles, all the same. This firm also used machinery for the entirety of their guns' manufacture, but took the process to a new level. Unlike Colt's revolvers, their rifles were not just visually the same, they were identical in that any part from one could replace the same part on another without any specialist fitting. They were truly interchangeable and, as such, were harbingers of momentous change in the history of gunmaking in Europe.

When the Pattern 1853 rifle was approved and manufacture about to commence, some remembered what they had seen at the Crystal Palace and what they had witnessed at Colt's factory. It potentially opened the doors to several opportunities for the government: to itself become a manufacturer in the true sense, and therefore remove the risk of being held to ransom in time of war if private contractors were unable to deliver through strikes or other labour problems; to be able to manufacture arms more quickly and cheaply without the need for highly specialized and costly craftsmen; and to expedite manufacture and repair by making the arms interchangeable. That subtle change in the degree of sophistication was accomplished by a combination of using machines entirely in the manufacturing process, along with an extensive set of gauges by means of which the size and often complex shape of each component could be checked to ensure it conformed exactly.

This was a system initially developed on a limited scale in France in 1777 by Honoré le Blanc, working under the instructions of General Gribeauval. Le Blanc's efforts extended only as far as lock manufacture, but were noticed by Thomas Jefferson who, following the War of Independence, became the USA's representative in Paris. Jefferson was sufficiently impressed to send at least one sample of le Blanc's 1777 Pattern musket to America with the hope that the system might be of interest. At this time, in the aftermath of revolution, America was isolated from what had been her usual source of supply of most manufactured goods, including firearms. The country needed arms to be able to defend its newly won independence but had a very limited supply of craftsmen capable of making them. There, le Blanc's ideas were pursued in earnest, and by the 1850s a position had been reached in which a whole new technology had been created, enabling guns to be made accurately and completely by machinery.

This became known as the 'American System of Manufacture'. The skill of the craftsman was replaced by the skill and ingenuity of machine-tool designers and builders and many new machines had to be developed to accomplish all the tasks. But it was not just the machines that were important – it was the way in which they were used. The system relied on two key features. One was the use of a single machine to carry out a single operation, rather than, as was usual in most workshops, having one machine which was continually re-set and adjusted to carry out a multitude of different machining operations. The other was the use of gauges to check both the dimensions and shapes of the finished components. If all the components passed the muster of the gauges, then they would fit together when it came to assemble the gun.

For Britain, the result was that the Royal Small Arms Manufactory at Enfield would start systematic manufacture of Pattern 1853 rifles, using purpose-designed machinery specially commissioned and imported from the United States. The first rifles began to appear in 1857. To give some idea of the scale of the operation, the making of a Pattern 1853 rifle, which has around 63 components (depending on the type of bands used) requires 719 separate machining operations carried out by 680 machines; this process was capable of producing 1,200 rifles per week. Such an operation involved a very large capital outlay beyond the means of the traditional contractors – only the government could afford to do it. In the year ending March 1858, 26,739 rifles were manufactured and a new era in gunmaking was under way. The machinery was installed and put into commission under the guidance of James Burton, who had gained practical experience of this technology in America. He later returned there and applied all he had learned at Enfield to making the same weapon for the Confederacy.

These methods, when applied to the manufacture of firearms, became known in America as 'armory practice' but it was quickly realized that the same technology could also be applied to the manufacture of other 'engineered' goods. One of the earliest to adopt this method was Aaron Dennison, who in 1850 founded the Waltham Watch Company and thus laid the foundations of the American watch and clock industry which was to outstrip even that of Switzerland, and which effectively destroyed Britain's watchmaking industries in Coventry and Prescott. Another eager practitioner around the same time was Isaac Singer, who produced what was probably the first commercially successful sewing machine. He also added another feature, the invention of the 'instalment purchase plan', which greatly expanded his market by making sewing machines accessible to almost every home. Even gunmakers began to see a wider potential. After the Civil War, when demand for guns fell, companies such as Remington began to diversify, in this case producing the first commercially successful typewriter. This technology, so obviously adaptable to producing a great variety of goods, along with the reduction in manufacturing costs that it allowed, made it the bedrock of the 'consumer society'.

Sadly, very few of the original machines used to make the Pattern 1853 survive, but a set of the gauges used at Enfield, many bearing the date of 1857, still exists. It is a very comprehensive set and most of the gauges are complex, reflecting not only the shapes of the different components, and in some cases assembled components, but also the range of different sizes of parts of those components.

A compound gauge, dated 1857, for checking the breech end of the barrel for correct position and profile of the bolster upon which the nipple fitted, and the size, shape and position of the tang which forms part of the breech plug, or pin, which screws into the end of the barrel. (Royal Armouries, ex-MoD Pattern Room Collection, PR.10142 [part])

SURGERY, MEDICINE AND HEALTH IN THE CRIMEAN WAR

Though it may be repugnant, one cannot overlook the fact that the purpose of any military firearm is to wound or kill an enemy. Just as the Crimean War took place in a period of revolution in military firearms design and technology, so too was it fought at a time when surgery had recently also undergone a revolution.

For centuries, surgery had been a nightmare of agony, but in 1846 and 1847 the almost miraculous anaesthetic properties of ether, and then chloroform, had been discovered. The soldier unlucky enough to be wounded at this time could at least hope for a reprieve from his pain if surgery was undertaken.

Battlefield surgeons were of course familiar with gunshot wounds, but up until this time they had been caused by round balls fired from smoothbore guns. It was quickly discovered that a rifle bullet behaved very differently from a round musket ball. In an extensive account of his experiences in the application of medicine and surgery at the Crimea, George Macleod, who practised his skills at Constantinople, Scutari, Smyrna and the hospital in the camp at Sebastopol, gives some insights into this.[34] 'But,' he says in his preface, 'this great war has, unfortunately, added little to our medical knowledge. Its short duration prevented this; yet it has shown us wounds of a severity, perhaps, never before equalled; it has enabled us to observe the effects of missiles introduced for the first time into warfare.' He later goes on to say:

> There was yet another element which demands attention, when estimating the surgical records of the war. I refer to the use of the new rifle, with its conical ball... The greater precision in aim, the immensely increased range, the peculiar shape, great force, and unwonted motion imparted by the new rifles to their conical balls have introduced into the prognosis of gunshot wounds an element of the utmost importance. I am not prepared to say whether the great destruction of the soft and hard tissues which these balls occasion results from their wedgelike shape, immense force and velocity, or the revolving motion, or from a combination of all of these causes together; but of one thing I am convinced, that their use has changed the bearing of many points which fall to be considered by the surgeon in the field.

It is obvious from this short passage that the use of the rifle on a large scale for the first time had a major effect on the way in which military surgery was practised. And of course central to this was the human body. Macleod gives further enlightenment on this also, as we would expect, though it does not make for comfortable reading:

> A round ball ... may be flattened against the shaft of a long bone, without causing any subsequent harm. This was often seen in India where the matchlock was used. The ball may be split into two parts on the edges of the tibia [shin bone] or the bridge of the nose. It may turn round a bone without breaking it. A round ball, as is well known, may notch or partly perforate a long bone [e.g., femur, or thigh bone] without causing fracture ... If the force of propulsion be a little greater, then the bone may be split longitudinally without being fractured across ... [The round ball] may penetrate into the spongy heads of bones and become encysted there. It may pass through, causing a clean hole ... but the conical ball never acts in any of these ways, so far as I have seen.

[34] All extracts below are taken from Macleod, George H. B., *Notes on the Surgery of the War in the Crimea*

In contrast, the conical round from a rifle

> ... is seldom split itself, but invariably splinters the bone against which it strikes to a greater or lesser degree ... but while a fracture with little comminution [shattering] results from a round ball, the conical ball – especially that which has a broad, deep cup in its base – splits and rends the bone so extensively that narrow fragments, many inches in length, are detached, and lesser portions are thrown in all directions, crosswise at the seat of the fracture, and driven into the neighbouring soft parts. It is the pressure of these fragments ... which renders the fracture of long bones by the new ball so hopeless.

We are beginning to see a picture of the greater damage to the body, even if we limit it to the limbs, caused by a conical bullet compared with a spherical ball. And, all else being equal, whereas the lead musket ball might not result in the need for amputation, from what we are told the conical bullet probably invariably did require such drastic measures. Not only that, but the bullet was inflicting worse injury at much greater range than the old ball. The same author tells us that, 'I have never met with an instance in which such a ball [he refers to the bullet], fired at whatever range, and striking at all perpendicularly on a long bone, has failed to traverse it and comminute it extensively.'

Musket balls have been known to meander through the body, sometimes doing surprisingly little major damage. But the conical ball, because of its higher velocity, the force of its impact and its more streamlined shape, 'seldom fails to take the shortest cut through a cavity or limb, and it has been at times seen, as at the Alma, to pass though the bodies of two men and lodge in the body of a third.'

A fallen Confederate soldier, with his Enfield rifle laid across him, near Spotsylvania Court House, Virginia, May 1864. (Library of Congress)

There can be little doubt, taking into account even the very few instances outlined, of the far greater effectiveness of the newer breed of rifle over the older smoothbore musket against human targets on the battlefield. And that, after all, was what it was all about. Unfortunately, the wound was not the only hazard faced by the unlucky soldier. While surgery had witnessed one miracle with anaesthesia, it still awaited the second – the ability to eradicate the appalling infection which often accompanied even the most minor surgery or slight wound. Anaesthetics were a mixed blessing. In addition to removing pain, they removed the need for speed, and this meant that the wounds were exposed for much longer periods and therefore put the patient at much greater risk of infection – often life-threatening – beyond that introduced by the dirty hands, instruments and clothing of a surgeon. In 1847 Philipp Ignaz Semmelweis had discovered that infection could be avoided by scrupulous cleanliness and the use of disinfectants for washing hands and instruments; however the old-school medical profession not only refused to accept Semmelweis's doctrine but ridiculed it, condemning thousands to an unnecessary, lingering, painful death. Another ten agonizing years were to pass after the Crimean War before Joseph Lister thrust antiseptic surgery on the world.

Surgeons in the Indian Mutiny and those in the American Civil War record the same effects, both in the nature of the wounds from rifles and in the treatments required. And perhaps the scale of mortality in the American Civil War can be attributed not only to the numbers of men engaged but also to the nature of the weapons being used. In some respects the surgical impact of the rifle bullet – whether from the French Minié, the British Enfield or the American Springfield – was the writing on the wall for future wars as sizes of bullets decreased and their velocities increased, with consequently even more devastating effects upon the human body.

Strange as it may seem, the field of surgery which probably derived most immediate benefit from the battlefields of the Crimea was gynaecology! At the outbreak of war, a young surgeon called Spencer Wells volunteered, and was sent first to Smyrna, where he was attached as surgeon to the British Civil Hospital, and afterwards to Renkioi in the Dardanelles. His experience with abdominal wounds in the Crimea showed him that the peritoneum, the membrane which encases all the organs in the abdominal cavity, was much more tolerant of injury than was generally supposed. After returning to London he utilized this knowledge to carry out his first ovariotomy in 1858. Although the patient died, he was not disheartened and devoted himself assiduously to successfully perfecting the technique, resulting in the saving of many women's lives throughout the world and pioneering the revolutionary field of abdominal surgery. In the process he invented the famous 'Spencer Wells' artery forceps still in use today.

TACTICS

There is little to suggest that the new rifle did much to change the way wars were fought. In fairness this is not surprising. At the time of the

Pattern 1853's first use in the Crimea, there had been little time to consider how it might change tactics and there was no experience of the use of the rifle on a large scale to provide any guidance. The Indian Mutiny seemed to have added little more, though in both conflicts there was an emerging understanding that this was a weapon capable of greater effect and at longer range than anything used previously. If anything, in the Crimea it probably highlighted the viability and value of long-range sharpshooting against strategic targets. Major Alfred Mordecai commented on these very points. He noted that:

> ... the ball, having a greater initial velocity, and a more advantageous shape ... describes a flatter trajectory, and retains a greater force of penetration at moderate ranges; and although the effect of the shot at great distances may be less than fatal than that of the larger calibre, it can hardly be doubted that a ball of the size adopted for our new rifle arms will be effective to disable a man or a horse at as great a distance as the limit of distinct vision can enable a soldier to fire with any accuracy.

He then went on to say:

> During the late war in Europe the great body of infantry ... used the ordinary musket, and this circumstance, combined with the limited scale of operations in the open field, did not allow a full trial of the effect of the new arms, and their influence on the tactics of armies.[35]

Both these attributes – long-range effect and accuracy, sometimes assisted by a telescopic sight – were actively embraced in the American Civil War. But, like the Crimean War, that war was also fought with a mixture of smoothbore muskets and rifles. Without a uniformity of firearms within an army it becomes very difficult, if not impossible, to develop and employ tactics better suited to the rifle.

Sharpshooting demanded an almost inherent skill alongside extensive training and practice, none of which, beyond the basic training, were likely to be available to the 'common soldier' – although as we have seen, the rise of the Volunteer regiments created a new 'class' of soldier and a new approach to this activity. Also, 'sharpshooting' by the ordinary soldier, as distinct from the Rifle Brigade, required the delegation of some independence of action – not something easily embraced by an army hierarchy which required that everything be 'done by numbers'! Many opportunities for different ways of conducting warfare offered by the use of the rifle were overlooked, or possibly not even seen by those in a position to bring about change. Thus, there was little difference from the way earlier wars had been fought – based on volley firing under the command of a sergeant and bayonet charges. It was a pattern to be repeated in France 50 years later, but against far more formidable weapons than any encountered before.

[35] Mordecai, A., Maj, *Military Commission to Europe in 1855 and 1856*

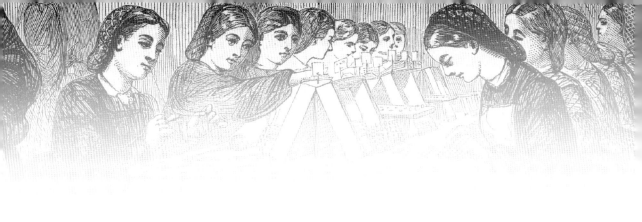

ACCESSORIES

A variety of items were officially issued with every rifle, including a ramrod, a bayonet and scabbard, a muzzle stopper, an oil bottle, a snap cap (nipple protector) with chain, and a nipple wrench along with spare nipples.

RAMROD

Although the ramrod was basically an integral part of the rifle, it is worth noting that two patterns were issued. The first had a mushroom head and a swell on the shaft, but during the Crimean War this was found to be problematical. Some time between 1857 and 1860, an alternative rod was introduced, which had a gradually tapering shaft and a 'jag' head. This jag head was a part of (or an attachment for) a ramrod, with serrations to grip the cloth that was wrapped around it for cleaning out the barrel.

The jag head on the second pattern allowed a better grip when it was being withdrawn from the ramrod channel beneath the barrel, and it also gripped any cloth patches used for cleaning the bore. The first pattern of rod would also have had a separate brass jag that screwed onto it when

The Pattern 1853 ramrods, first pattern (top) and second pattern (bottom). The slot in the jag was to allow the screwdriver blade on the nipple wrench to be inserted to give extra torque when required. (Private collection)

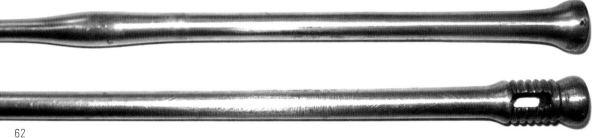

required, but no example has been located. The tip of the head of each was hollowed to match the contour of the bullet nose so that during loading the nose of the bullet would not be distorted when being pushed down the bore.

Before the Enfield factory got into full production and for some time afterwards, ramrods, like other components of the rifles, were made by contractors. Usually the name or initials of the contractor can be found stamped near the head, along with government inspector's marks. Also, on many ramrods that are clean but have become tarnished, a diagonal line can often be seen cutting across the shaft a few inches from the head. This shows the scarf joint where the steel and iron of shaft and head were forge-welded together.

BAYONET

The bayonet, like the ramrod, was not so much an accessory as an integral part of the weapon, and the standard 1853 Pattern socket bayonet was designed at the same time as the rifle. It had a browned (blued) socket fitted with a locking ring, and a polished cast-steel blade of equilateral triangular section, fullered (hollow ground) on each face and forge-welded to the iron socket. It was accompanied by a black leather scabbard with brass locket (mouthpiece) and chape (tip). Two patterns of scabbard were issued; the first pattern had the 'frog stud' fixed directly to the leather, but in 1860 this was changed to being fitted at the bottom of an extended locket instead. The 'frog' was the leather sheath in which the scabbard was fitted for carrying on a waist belt.

Bayonets and scabbards were allocated a service life of six years, after which time they were returned to store and replacements issued. In actual fact, another difference exists which is not always easily apparent. After 1857 the rifled musket was manufactured so as to have interchangeable components, and therefore conformed to rigid dimensions. The same applied to the bayonet, so any bayonet will, theoretically, fit any Enfield-made rifle. Prior to that, bayonets were manufactured by contractors; they were selected to fit a particular rifle and were identified accordingly. Great care was taken to ensure that the pair remained together during service, and if they were to be shipped in arms chests or put into store they were actually tied together with string.

Pattern 1853 bayonet and scabbards. Upper scabbard is the first pattern, lower the second pattern, showing the different ways in which the frog studs were mounted. (Royal Armouries, ex-MoD Pattern Room Collection, PR.1913 and PR.1914)

Snap cap in place on the nipple and secured to the rifle by a chain. On some weapons there was a special small screw-eye through the front trigger guard tang to which the chain was attached. (Royal Armouries XII.1365)

TOMPION, OIL BOTTLE AND SNAP CAP

The tompion, or muzzle stopper, was a simple item designed to prevent the ingress of foreign matter into the bore of the rifle when not in use. It usually consisted of a stack of cork discs about 1in long on an iron pin, held in place with a large washer at the bottom and the pin secured in a brass top.

The oil bottle was a simple device consisting of a cylindrical zinc body and base, 2½in tall with a diameter of 1in, fitted with a brass top with a screw-on brass stopper. The stopper was fitted with a long spatula-like pin used to transfer drops of oil to the various lubrication points. The oil itself was specified as Rangoon Oil and the bottle was carried in the ammunition 'ball bag' (not in the pouch, which was a separate item).

The snap cap was a device designed to fit onto the square of the nipple; it had a cavity on its upper side fitted with a thick, well-oiled leather washer on which the hammer nose rested, off the face of the nipple, protecting both from damage. This was especially important, not just when the rifles were not in use, but also during firing drill when the trigger was pulled with the hammer at full cock. Also, with a snap cap and tompion in place, the interior of the barrel was sealed and, in theory, prevented any ingress of moisture.

NIPPLES AND NIPPLE WRENCH

Nipples were a vital part of the functional firearm, but were subject to breakage on occasions and to erosion from the hot flame from exploding fulminate. Obviously in service, a rifle with a faulty nipple is useless and spares were standard issue. The nipple, made of hardened steel, was a precisely manufactured item.

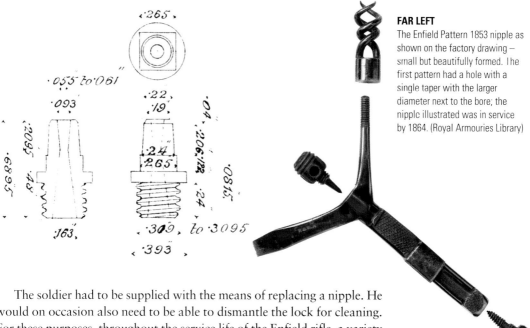

FAR LEFT
The Enfield Pattern 1853 nipple as shown on the factory drawing – small but beautifully formed. The first pattern had a hole with a single taper with the larger diameter next to the bore; the nipple illustrated was in service by 1864. (Royal Armouries Library)

ABOVE
Sergeants' Pattern No. 1 'Y' nipple wrench with the initials of the contractor (R & W Aston) on the turnscrew blade. With its special intense blue finish, this is probably one of the finest examples of a nipple wrench likely to be encountered and forms part of the cased rifle and accessories presented to Lord Panmure, Secretary of State for War (1855–58) under Lord Palmerston. (Royal Armouries XII.10598)

The soldier had to be supplied with the means of replacing a nipple. He would on occasion also need to be able to dismantle the lock for cleaning. For these purposes, throughout the service life of the Enfield rifle, a variety of nipple wrenches, or 'combination tools', were issued, the first being approved in 1855, the same year that general issue of the rifle began. Usually two types of whichever pattern was used existed; one was for sergeants and the other for privates.

The Pattern No. 1 was of the so-called 'tri-arm' type, i.e. 'Y'-shaped. There were major differences between the sergeants' and privates' versions. Both had a screwdriver blade, a worm, an oil bottle formed in the hollow cavity within the main body, and a square socket at one end to fit the nipple, but the sergeants' version also carried a mainspring clamp, a drift and a ball-drawer. In the privates' version, the screw-on stopper for the oil bottle, fitted with a leather washer to prevent oil leakage, carried a long 'oil-pin' with which to apply drops of oil at strategic places. In the sergeants' version the drift also formed the cap of the oil bottle while the ball-drawer was screwed into a hole formed at the bottom of the square socket which fitted the nipple. In use, the drift was attached to the ramrod and, by working the ramrod up and down, was used to form a tapered hole in the nose of a bullet lodged in the bore. The drift was then replaced on the ramrod by the ball-drawer, shaped like a woodscrew, which was then screwed into the hole formed by the drift and the bullet pulled out. To aid these processes, the slot in the head of the second pattern ramrod was made to allow the large screwdriver blade to act as a lever. Most, if not all, of these differences continued through the remaining series of nipple wrenches.

Pattern No. 2 was for sergeants only and was produced experimentally in early 1858. It was very similar to Pattern No. 1 but the arm carrying the worm was pivoted so that it could be folded away. It was never adopted and is extremely rare.

Pattern No. 3 was the first of the 'T'-handled wrenches for sergeants and privates, and production began in late 1858. Unlike the 'Y'-shaped versions this was made in two parts, consisting of the body and the crosspiece. In the first examples, the end of the body was formed into a square so as to fit a corresponding hole in the crosspiece. However, this was discovered to be too weak and was replaced by a teardrop-sectioned projection on the body fitting into a matching hole on the crosspiece. It was this second version which received final approval as the sealed pattern. In other respects it was the same as the Pattern No. 1 except that the cap to the oil bottle also held the crosspiece in place on the body.

Pattern No. 4 appeared in 1859 and was simply a slightly smaller version of No 3.

Pattern No. 5 resulted from the change to the Baddeley barrel bands in 1861, which required a smaller screwdriver to use with the deeply countersunk screws. It was basically the No. 3 Pattern but with the tip of the thread carrying the worm formed into a small screwdriver blade.

The final version, which logically should perhaps be Pattern No. 6, but is referred to as Pattern No. 7, appeared in 1862 and simply gave the sergeants' and privates' wrenches greater consistency by making the prickers (the needle-like tools used for cleaning the vent in the nipple) the same length, and making the privates' oil bottle cap the same form as that of the sergeants' version. The need for a nipple wrench ceased with the introduction of the Snider breech-loading rifle in 1866 and they were gradually withdrawn from service.

ARMOURER'S TOOLS

These are not strictly speaking accessories in the usual sense, in that they were issued with each rifle, but they were certainly an important accompaniment to regimental issue. Whereas the nipple wrench allowed privates to carry out a certain level of maintenance and repair, and sergeants an additional level, a point would be reached when the rifle or other arms required the skills of a specialist craftsman, in this case the armourer sergeant. The 'armourer's forge', as the set of equipment was called, in 1859 comprised 173 tools of great variety, since the armourer was not only entrusted with the care and repair of rifles, but had all manner of other duties thrust upon him when the skills of a craftsman were required. And he really did possess a forge with the appurtenances one would associate with that – bellows, anvil, tongs, hammers, etc. But it would be impossible, and in a great many cases irrelevant, to give a detailed listing. As far as the rifle was concerned, he had such items as gauges, gouges, files and floats for work on the stock; grinders, screw plates and taps for making or adjusting the variety of screws and screw holes encountered; hand brace and drills for making holes in metal and wood; and wrenches for the breech-pin. Today the 'grinders' would probably be called something like 'hollow end mills' or 'counterbores' or 'countersinks'. To add confusion, the term 'nail' was the official designation for the side and breech tang screws, while the screws of the lock were referred to as 'pins', but the screws of the furniture and bands were in fact called 'screws'.

FIRING THE ENFIELD PATTERN 1853

The soldier used pre-prepared cartridges carried in an ammunition pouch. By around 1864 at least four varieties of pouches were in use, two specially for the rank-and-file and sergeants of Guards, and two for the other infantry regiments. It was never British military practice to load powder directly from a flask.

As specified in *Instruction of Musketry*, the cartridges themselves consisted of the standard lead bullet, loaded so that its nose pointed inwards, and 2.5 drams of Enfield rifle (ER) powder made at the government's gunpowder mills at Waltham Abbey, all contained in a paper cylinder. Once the cartridge had been made up, it was warmed and the bullet end was dipped vertically up to the shoulder of the bullet in a pot of molten lubricant.

As already mentioned, during the 1852 experiments the original Minié design of bullet was used with the result that, because it was entirely conoidal, it did not always align correctly in the bore and thus produced inaccurate shooting.

Pritchett's bullet, adapted from William Metford's design, was adopted for use in the .577cal rifle. Although it did not require a base plug to assist with expansion into the rifling grooves – if accurately made to the diameter specified – the expedient of an iron cup, along with a flat bottom to the cavity, was adopted by the Royal Laboratory in its haste to manufacture cartridges for the troops in the Crimea. This was then changed, and wooden and baked-clay plugs were used.

A feature of government-made bullets, often overlooked by collectors and enthusiasts, is that

BELOW LEFT
A typical Enfield cartridge. The staining around the bullet portion at the bottom is from the lubricant. It is believed this particular cartridge may be one of those prepared during the 1852 trials before final adoption in 1853. (Private collection)

BELOW RIGHT
A standard cartridge in section, showing the bullet with its wooden plug, nose-inward. (Royal Armouries, ex-MoD Pattern Room Collection)

ABOVE
Nose of the punch used to create the flat-bottomed base cavity, with its four broad arrows cut into the rim so that the finished bullet has these reproduced in embossed form. (Royal Armouries, ex-MoD Pattern Room Collection)

ABOVE RIGHT
Enfield bullet in unfired condition recovered from the American Civil War battlefield of Vicksburg, showing in the cavity the manufacturer's code '57' for Eley Brothers. (Private collection)

on the rim of the skirt are four embossed broad arrows, the mark of government ownership. These are placed there in one of the final bullet-making operations by the punch that creates the cavity in the base. Also, on the flat crown of the cavity a number is usually stamped, coded to indicate the manufacturer.

It is also worth noting that British service bullets were round-nosed and did not have cannelures, unlike some modern replicas which have both cannelures and a flat nose. There were actually two different bullets used. Initially, a bullet of 0.568in diameter and 1.05in long was the standard. However, following recommendations put forward in 1857 by Captain Boxer, RA, Superintendent of the Royal Laboratory, on 26 July 1858 a lubricant consisting of five parts beeswax and one part tallow was adopted for service in India and then, in early 1859, beeswax alone was adopted throughout the whole of the British Army. Because beeswax was a harder lubricant, a reduction of bullet diameter to 0.55in was implemented to maintain ease of loading. At the same time, the bullet length was increased to 1.09in. This slight increase in windage was reported not to have any effect on the accuracy of shooting and the anti-fouling properties were greatly improved over lubricants containing tallow. These bullets were not made by casting, which, among other possible defects, often causes entrainment of small air bubbles, whose quantity and distribution creates variability in ballistic properties. By pressing them in dies, from a slug of extruded lead wire, this problem was largely eliminated, producing bullets of uniform but higher density, both of which are essential qualities if consistent firing results are to be obtained.

Cartridges were manufactured at the Royal Arsenal, Woolwich, and also by private contractors in very large numbers. In 1862, for instance, we are told by the *Illustrated London News* that at Woolwich the output was around 900,000 per week. When the bullets began to be made by machine, it became a predominantly manual process, the first stage consisting of rolling the paper on the former, inserting a bullet and tying the end off with tightly knotted fine string or thread.

Young girls at work preparing the empty cartridge tubes at Woolwich. They worked a five-day week of nine and a half hours per day with time off for breakfast and lunch. (*Illustrated London News*, 1862/Private collection)

This stage of manufacture was carried out by 200 girls, who were daughters of either soldiers or arsenal workmen, aged between ten and 14 years. Working on a piecework basis, the 'nimbleness with which their little fingers rattle over the work' was sufficient to earn them between eight and 11 shillings (40–55p) per week, a considerable amount at that time, especially for children. And unlike many other children employed in factories, they were particularly well cared for. After the empty tubes had been prepared, they were taken to the magazine for 'powdering'.

Of course, in extreme circumstances soldiers sometimes had to 'make do and mend', creating their own cartridges and bullets. To this end, instructions were supplied along with sets of tools. The shapes and sizes of the paper used were carefully defined, as shown in the accompanying diagram, and to aid cutting out a set of tin templates was provided.

The paper itself was of a special type, patented by John Dickinson in 1807; in addition to cotton fibres, it contained a small proportion of wool which prevented smouldering after initial ignition. This was important in that it theoretically ensured that after the gun had been fired, any paper residue left in the bore would not be in a state to ignite the charge of powder poured in for the next shot.

The other essential item for firing was of course the percussion cap and these were also prepared at the Waltham Abbey factory. They were very simply made from a sheet of copper, by stamping out cruciform-shape pieces which were then pressed in a die to form them into a cup shape, followed by folding the tips of the cross outwards to give the characteristic 'top-hat' shape. The machine used for carrying out this series of operations was capable of making 130,000 empty caps in a ten-hour day.

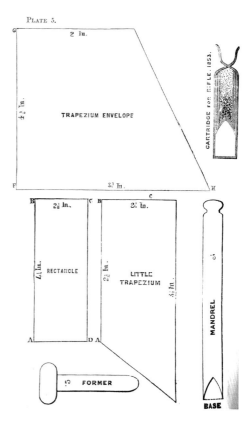

Templates and tools used in making up the Enfield Pattern 1853 cartridge as shown in *Instruction of Musketry*, 1854. (Royal Armouries Library)

The caps now had to have the fulminate added and a combination of machinery and the human hand was used to perform this task. It is perhaps worth bearing in mind at this point that, such was the extremely sensitive and explosive nature of the fulminate, that only 12oz was allowed in total in the factory at any one time! One thousand caps were placed in holes on a brass tray. A second plate with the same number and spacing of the holes was placed over the plate carrying the caps but with the holes offset. Onto this plate was placed a quantity of fulminate, which was spread around with a cardboard spreader until all the holes were filled. The plate was of such a thickness that each cavity held one-third of a grain. When the excess fulminate had been removed, the plates were moved using a fine screw, to avoid rapid movement of the plates together and frictional detonation of any trace of fulminate trapped between them, until the cavities containing the fulminate coincided with the caps on the plate beneath. The correct quantity of fulminate was thus deposited in the caps.

The tray of filled caps was then taken and fitted into the pressing machine. As the machine was operated, a ratchet arrangement moved the tray forward one row at a time and a row of punches descended to compact the fulminate in the head of the cap. When completed, the tray of caps was then transferred to a varnishing machine which operated in a similar manner to the pressing machine, moving the tray forward one row at a time. As each row came into position, the boy controlling the machine swung an overhead bar containing long pins fitted so that they coincided with the caps in the tray. These pins were first dipped into a solution of shellac in alcohol, then moved so that they were positioned over the caps. This movement was controlled by linkages so that the positioning was exact. The pins were then lowered into the caps and the drop of varnish on the tip of each was taken up by the caps and was sufficient, after drying, both to secure the fulminate in place and to render it waterproof.

When issued from store, the ammunition was packed either in quarter barrels or in boxes. Quarter barrels were 14¼in long with a diameter of 11⅝in and contained 700 cartridges in waterproof bags and 1,050 caps in a zinc cylinder. The top of the barrel was painted black and a white printed label detailing the contents was attached.

For all stations in the tropics or for service in the field, as well as for China, boxes were used and, for the tropics, were made of teak with mahogany ends. Boxes were 16½in long, 7¼in wide and 8½in deep. Initially the boxes could accommodate 560 cartridges and 700 caps in a zinc box. However, with the adoption of waterproof bags, there was only room for 440 cartridges and 660 caps.

When it came to actual military use of the rifle, its loading and firing followed very precise procedures laid down in the *Infantry Manual*. The actual procedures depended upon whether the soldier was standing, front-rank kneeling or rear-rank kneeling, and this mainly concerned the bringing of the rifle into a suitable position for loading. However, the actual loading and firing process itself followed a common route.

When standing, the rifle was held vertically with the left hand, 6in from the body and with the top of the barrel facing forward.

The instructions then continue:

1st. Bring the cartridge to the mouth, holding it between forefinger and thumb, with the ball in the hand, and bite off the top, elbow close to the body.
2nd. Raise the elbow square with the shoulder, with the palm of the hand inclined to the front, and shake the powder into the barrel.
3rd. Reverse the cartridge (keeping the elbow square) by dropping the hand over the muzzle, the fingers in front of the barrel, and place the bullet into the barrel nearly as far as the top, holding the paper above it, between forefinger and thumb.
4th. By a turn of the wrist from left to right, tear off the paper that remains between forefinger and thumb, dropping the elbow to the side at the same time, and seize the head of the ramrod with the second joint of the forefinger and thumb.
5th. Force the ramrod half out, and seize it backhanded exactly in the middle, the elbow square with the shoulder.
6th. Draw it entirely out with a straight arm above the shoulder, turning it at the same time to the front, put it on top of the bullet, turning the back of the hand to the front; the ramrod is thus held between the two forefingers and thumb, with the last two fingers shut in the hand.
7th. Force the bullet straight down till the second finger of the right hand touches the muzzle; elbow close.
8th. Press the ramrod slightly towards you, and slip the two forefingers and thumb to the point and grasp it as before.
9th. Force the bullet steadily straight down to the bottom, bringing the elbow down with it close into the body.
10th. Ascertain that the bullet is resting on the powder by two slight taps, avoiding all sharp strokes.
11th. Draw the ramrod half out, catching it backhanded, with the elbow square.
12th. Draw it entirely out with a straight arm above the shoulder, turning it to the front; put it into the loops, and force it as quickly as possible to the bottom, the forefinger and thumb holding the ramrod as in the position immediately previous to drawing it.

The barrel is now charged but capping is still required:

… bring the firelock at the same time to the right side, with the left hand at the swell, and hold it in a horizontal position … the right hand holding the small of the butt, and half cock the piece, the thumb resting on the cock.

Advance the forefinger to throw off the old cap (after having fired) Carry the hand to the cap pocket, and take a cap between forefinger and thumb.

Put the cap on the nipple, and press it down with the flat part of the thumb; fingers clenched. Then carry the hand to the small of the butt.[36]

[36] *The Infantry Manual, 1854*

FIRING THE PATTERN 1853

1. At the ready
2–3. Cartridges were issued in string-wrapped paper packets
4. Cartridge withdrawn from the pouch
5. After tearing off the top of the cartridge with the teeth, the charge of powder is poured down the barrel
6. The cartridge is reversed, the bullet with its paper wrapping entered into the muzzle and the excess paper torn away
7. The ramrod is withdrawn, inverted, and used to push the bullet down the bore so that it sits firmly on the charge of powder

8. If there is an appreciable gap between bullet and powder, the barrel is likely to burst when fired
9–10. After pulling the hammer back to half-cocked position, a percussion cap is placed upon the nipple and pressed firmly into place with the thumb
11. After making any sight adjustments the rifle is fully cocked, brought up to the shoulder and aimed
12. Firing

The rifle was finally fully cocked after adjusting the sight and shouldered for firing.

All of these manipulations were accompanied by other evolutions, such as correct positioning of the body for rank firing. Notice the use of the term 'firelock', repeated elsewhere in the document, which seems completely archaic and out of place, especially when the document specifically refers to the rifle musket, which, it must be remembered, was a highly refined piece of equipment. More importantly, however, it will have been noted that it was *not* the bullet end of the cartridge that was bitten, contrary to popular view. This raises the obvious question of why the Indian native troops objected, since only the bullet was greased. However, another element of firing procedure was that 'whenever the lubricating material round the bullet appears to be melted away, or otherwise removed from the cartridge, the soldier is to be instructed that the sides of the bullet should be wetted in the mouth before putting into the barrel, as the saliva will serve the purpose of grease for the time being'.[37] Considering the climate in India, this may have had some bearing on what happened there!

The series of photographs on the previous pages show the loading and firing sequence as practised by the Confederate army, and were specially demonstrated for this book by David Naumec of the American North–South Skirmish Association and photographed by fellow member Dean Nelson, both of whom are museum curators in Connecticut. The inclusion of Confederate practice is appropriate since the British pattern of Enfield cartridge was the standard adopted by General Gorgas and closely followed British military practice.

It was generally claimed that, with practice, a rifleman could load and fire the Pattern 1853 rifle three times in one minute. Under the stress of battle, however, one wonders how many of the small caps slipped out of sweaty, or cold-numbed, or shaking, greasy fingers; or how many bullets were loaded the wrong way.

Despite the cumbersome wording of these firing procedures, much the same process is followed by today's users of these rifles, varying perhaps according to whether they are being used on the range or as part of some re-enactment. In the past most users on the range tended to load from a flask, but current regulations now require the use of pre-prepared charges, though these do not generally take the form of a paper cartridge. Another notable difference is that many modern bullets that have been examined by the author do not comply with the regulation type or sizes; many have cannelures; many are flat-nosed; many are oversized and greater force is required to load them than the two fingers and thumb indicated in the regulation procedures. This possibly arises from the fact that many bullets are cast in two-part moulds which often makes them slightly asymmetrical anyway, and are not then properly sized in a die. Some have even been measured at over 0.577in!

---

[37] *Instruction of Musketry*, 1856

CONCLUSION

The middle years of the 19th century were a period of unprecedented revolution. The Pattern 1853 rifle musket was itself conceived from the revolutionary ideas of Claude Etienne Minié. It was innovative in its design, abandoning preconceived ideas of how a British musket should be constructed, and it is one of those rare examples of a hybrid designed by committee achieving perfection. It revolutionized the British Army, for the first time providing every soldier, rather than a select few, with a high-quality, accurate and long-range rifle, and a rigorous training regime to go with it. The establishment of the factory at Enfield to make it also revolutionized the government's weapons procurement process. It, and its brethren, might also be said to have revolutionized the sporting world in creating long-range target shooting as a pastime as well as a military discipline.

Because of the visible effectiveness of this rifle in the Crimean War, especially its ability to outrange the Russian artillery, it precipitated a reappraisal of and then a revolution in artillery design, construction and manufacture. It, or more accurately its ammunition, played a part in tipping the scales of a revolt which we know as the Indian Mutiny or the Great Rebellion.

Wherever the British Army had a presence, so too did the Pattern 1853 and in that sense, it was international. Even where the British Army had no presence, the Enfield had its place and it found itself being used in large numbers by both the Union and the Confederacy in the American Civil War.

The quality and effectiveness of this rifle cannot be better expressed than in the words of Colonel Hay, Commandant of the School of Musketry at Hythe, in his Annual Report of 1856:

> Should there exist any one who retains any lingering affection for the musket pattern 1842, I would point attention to the trial ... where it will be seen that the best shot of this establishment, firing with that

musket from a rest, could not, at 300 yards, hit an 18 feet square target once in 20 shots taking aim at a mark in its centre, and that at 200 yards the shooting was hardly more effective; whereas at 800 and 1000 yards the same target will hardly ever be missed by a good shot firing with the rifle musket 1853. I have seen a target hit 96 times out of 100 successive shots, at 800 and 1000 yards, with the same rifle musket without cleaning.

When such facts as these stand officially recorded, the greatest admirer of 'Brown Bess' will be constrained to acknowledge and admire the foresight and firmness which, unshaken by all the argument which prejudice could suggest against the reform, have rescued our infantry from so useless a weapon as the musket 1842, and substituted in its place the most effective arm ever placed in the hands of a soldier, viz., the rifle musket 1853.

Nevertheless, it was soon to fall victim to another revolution in firearms technology – breech-loading. In 1864 a committee was established to explore ways of applying this technology to the Pattern 1853 rifle. Nearly 50 different methods of achieving this were proposed by gunmakers and inventors but it was the method proposed by Jacob Snider that was ultimately adopted. In 1866 conversion of all the Pattern 1853 rifles commenced. But the Pattern 1853 rifle did not really die; it was simply reborn as the 'Snider'.

The Pattern 1853 Enfield rifle musket has, therefore, numerous threads connecting it to many facets of human history and endeavour, but perhaps its greatest hidden legacy, still with us today, lies in the fact that it was the product of a revolutionary manufacturing technology imported from America. It was the first military firearm in Europe to be mass-produced with interchangeable parts, entirely by machinery. But this same technology, adopted and adapted by other manufacturers to mass-produce an ever-widening range of other types of goods, makes it representative of the beginnings of the consumer society.

GLOSSARY

BROWNED (BLUED) — the chemical treatment of gun barrels to help protect them from corrosion – usually resulting in a colour which could range from brown to almost black

CANNELURES — grooves formed in the periphery of cylindrical portion of the bullet

FLINTLOCK — ignition system by which a piece of flint was struck against a piece of hardened steel, thus generating a spark at the instant of firing

FOULING — residue remaining in the barrel after firing black powder (gunpowder)

FULLERED — the forming of a shallow trough along part of the length of a blade

FURNITURE — items which provide the 'finishing touches' to a gun – such as the butt plate; trigger guard; sideplate or cups which form a seating for the heads of the lock screws; fore-end cap

HOLLOW GROUND — giving a blade a concave face extending, more or less, its full length. This type of finish was usually only carried out on triangular section blades and was generally applied to two or all faces

MATCHLOCK — the oldest and simplest of firearm technologies, in which ignition was achieved when a length of smouldering twine or matchcord was thrust into a pan of priming powder

NIPPLE — a hollow pillar fitted to the breech of a gun, on which was placed a small copper cap containing fulminate, which was then struck to fire the gun

PERCUSSION — method of firing by which a cap containing fulminate was struck, and the flash from the resulting detonation passed down the hollow stem of the nipple and ignited the main charge in the barrel

WHEELLOCK — the first self-contained ignition system, in which pulling the trigger released a wheel which, driven by a spring, rotated rapidly against a piece of pyrites, throwing off a shower of sparks which ignited the powder in the priming pan; the resultant flame ignited the main charge in the barrel

WINDAGE — a small gap between the ball and the inside of the barrel in a muzzle-loading gun which eased loading but allowed some of the explosion gases to escape around the edges of the ball

BIBLIOGRAPHY

Atlas of the American Civil War, Oxford: Oxford University Press, 2004

Blackmore, Howard L., *British Military Firearms 1650–1850*, London: Herbert Jenkins, 1961

Bond, H., Lt-Col, *Treatise on Military Small Arms and Ammunition*, London: HMSO, 1884

Bowen, James L., *History of the Thirty-Seventh Regiment, Mass. Volunteers*, Holyoke, MA, and New York: Bryan & Co., 1884

Bradwell, G., *Under the Southern Cross: Soldier Life with Gordon Bradwell and the 31st Georgia* (compiled and edited by Pharris Deloach Johnson), Macon, GA: Mercer University Press, 1999

Bruce, George A., Brev Lt-Col, *The Twentieth Regiment of Massachusetts Volunteer Infantry, 1861–1865*, Boston, MA, and New York: Houghton, Mifflin & Co., 1906

Busk, H., *The Rifle and How to Use It*, London and New York: Routledge, Warnes & Routledge, 1859. Reprinted Surrey: Richmond Publishing Co., 1971

'Carman Manuscript', Antietam National Battlefield Library

Castel, Albert, ed., 'The War Album of Henry Dwight', *Civil War Times Illustrated*, May 1980

David, Saul, *The Indian Mutiny: 1857*, London: Penguin Books, 2003

Dupuy & Dupuy, *Encyclopaedia of Military History*, London: Macdonald & Co. Ltd, 1970

Glatthaar, Joseph T., *General Lee's Army: From Victory To Collapse*, New York: Free Press, 2008

Hawes, Arthur B., Capt, Bengal Army, *Rifle Ammunition – Notes on the Manufactures Connected Therewith at the Royal Arsenal, Woolwich*, London: W. O. Mitchell, 1859. Reprinted Gettysburg, GA: Thomas Publications, 2004

Hay, Charles, Colonel Commandant, *Annual Report of the Instruction and Experiment Which have Been Carried on at the School of Musketry, for the Year Ending 31st March 1856*, London: HMSO, 1856

Hess, Earl J., *The Rifle Musket in Civil War Combat: Reality and Myth*, Lawrence, KS: University Press of Kansas, 2008

Hibbert, Christopher, *The Great Mutiny: India 1857*, London: Allen Lane, 1980

Hope, A. R., *Story of the Indian Mutiny*, London: The London Printing and Publishing Company, 1860 (London: Warne & Co., 1896)

The Infantry Manual, London: War Office, 1854

Instruction of Musketry, London: HMSO, 1854

Lang, Arthur Moffatt, *Lahore to Lucknow: The Indian Mutiny Journal of Arthur Moffatt Lang*, London: Leo Cooper, 1992

'Lieut. Colonel Davidson's Patent Telescopic Rifle Sight', *Army and Navy Journal*, August 1864

Lincoln, William S., *Life with the Thirty Fourth Mass. Infantry in the War of the Rebellion*, Worcester, MA: Noyes, Snow & Co., 1879

Macleod, George H. B., *Notes on the Surgery of the War in the Crimea: With Remarks on the Treatment of Gunshot Wounds*, London: John Churchill; Philadelphia, PA: J. B. Lippincott & Co, 1862. Reprinted Kessinger Publishing's Legacy Reprints

McAulay, John D., *Civil War Small Arms of the U.S. Navy and Marine Corps*, Lincoln, RI: Andrew Mowbray Publishers, 1999

Military Commission to Europe, 1855 and 1856; report of Major Alfred Mordecai; US Ordnance Department: Washington DC, 1860

Musketry Instruction of the Army, London: HMSO, 1859

Newell, Joseph K., Capt, *Annals of 10th Regiment*, Springfield, MA: Nichols & Co., 1875

North British Review, February and May, 1858, Vol. XXVIII, American Edition Vol. XXIII, New York: Leonard Scott & Co., 1858; p.282

Pegler, M. *Out of Nowhere – A History of the Military Sniper*, Oxford: Osprey, 2004

Petrie, M., Capt, *Equipment of Infantry*, London: HMSO, 1863

Ray, Fred L., *Shock Troops of the Confederacy*, Ashville, NC: CFS Press, 2006

Report of Experiments with Small Arms carried on at The Royal Manufactory, Enfield, London: HMSO, 1852

Roads, C. H., *The British Soldier's Firearm, 1850–1864*, London: Herbert Jenkins, 1964

Roe, Alfred S., *The Twenty Fourth Regiment, Massachusetts Volunteers, 1861–1866*, Worcester, MA: Twenty Fourth Veteran Association, 1907

Russell, W. H., *The War: From the Landing at Gallipoli to the Death of Lord Raglan*, London and New York: Routledge, 1855

Smithurst, P. G., *Mallet's Mortars: A Great Experiment in Artillery*, Royal Armouries Yearbook, 2, 1997

Sword, Wiley, *Firepower From Abroad – The Confederate Enfield and The LeMat Revolver*, Lincoln, RI: Andrew Mowbray Publishers, 1986

Tennent, Sir J. Emerson, *The Story of the Guns*, London, Longman Green, 1864 (reprinted Richmond Publishing Co. (nd), Surrey)

Thorburn, S. S., *The Punjab in Peace and War*, London: William Blackwood and Sons, 1904

Thorwald, Jurgen, *The Century of the Surgeon*, New York: Pantheon Books, 1956

Tunnard, W. H., *A Southern Record: The History of the Third Regiment Louisiana Infantry*, based on his personal diary first printed in Baton Rouge, Louisiana for the author in 1866. Fayetteville, AR: University of Arkansas Press, 1997

Charles Usherwood's Service Journal, 1852–1856 (unpublished), Green Howards Regimental Museum, Richmond, North Yorkshire

INDEX

References to illustrations are shown in **bold**.

accessories 62–66; armourer's tools 66; nipple wrenches **65**, 65–66; nipples 64–66, **65**; oil bottle 64; ramrods 12, **62**, 62–63; snap cap **64**, 64; tompion 64; *see also* bayonets
Alma, battle of the (1854) 11, 26, **27**, 59
American Civil War (1861–65) 5, 39–41, **44**, 44–49, **45**, **46**, **47**, 59, 60, 61, 75
Antietam, battle of (1862) 46, **47**

'Baddeley' barrel bands 17, **18**, 19, 21, 66
Baker, Ezekiel 6; rifle **6**, 6
Balaclava **25**, 26; battle of (1854) 27–29, **28**, 33, 34
Barrackpore (now Barrackpur) 35
barrel bands, 'screw/spring' 17, **18**, 19, 21, **26**, 66
bayonets **15**, 15, **16**, **25**, **63**, 63; Pattern 1853 artillery carbine **19**, 19, 20
Birmingham, private arms trade 41, 55
Bitpur, Dandu Panth, Rajah of (Nana Sahib) 36
breech-loading 6, 6–7, 23, 76
British Army 11, 75; Durham Light Infantry, 68th **25**, 26; 'Green Howards' 26, **27**, 29, **32**; Guards **27**; Highlanders **37**; Highlanders, 93rd **28**, 28, 29; Light Brigade 27, 34; Light Division 26, **27**, 29, **32**; Regiment, 60th 21; Regiment, 93rd **37**, 38; Rifle Brigade 10, **25**, 25, **32**, 33; Rifle Corps 21; Rifle Regiment, 95th 6; Sergeants of the Line 21; Victoria Rifles 53
Brunswick rifle **8**, 8
bullet punch **68**, 68
bullet tests 14
bullets **12**, **14**, 14, 67–68, **68**; Minié **11**, 11, **12**, 14, 67; Pattern 1842 **12**
Burton, James 45, 57

Canadian Rifles, Royal 21
Cape Colony 8–10; Cape Mounted Rifles 21
caps, percussion, manufacture of 69–70
carbines 18–20; artillery **18**, 18–19, **19**; cavalry 20, **20**, **21**; Sappers and Miners 22–23
cartridges, paper 34–35, 36, 39, 45, 47, 48, 54, **67**, 67, 74; manufacture 68–69, **69**; 'Yankee' 47
Cawnpore: massacres **36**, 36, 37; relief of (1857) **36**, **37**
Colt, Samuel 55, 56
Confederate Army 44, 45, 46, 47, 75; Georgia, 2nd and 17th 48; Louisiana, 3rd 48; Mississippi Regiment, 2nd 40; soldiers **47**, **59**; South Carolina, 10th 44–45
Crimean War (1854–56) 10, 11, 24–29, **25**, **26**, **27**, **28**, **32**, 32–34, 54, 55, 61, 75

Davidson, Lt-Col D. **32**, 33
Delvigne, Capt Henri-Gustave 8
Dennison, Aaron 57
development 9–23; carbines **18**, 18–20, **19**, **20**, **21**; Indian service 21–22, 54; and Lancaster 22–23; rifle muskets 17–18; short rifles **21**, 21, 23; and Whitworth 22, 23
Dickinson, John 69
drill instruction 50, **51**, 51

Dum-Dum arsenal 34–35

Enfield, Royal Small Arms Manufactory 45, 56, 75

Ferguson, Patrick, and rifle **7**, 7
Forsyth, Rev Alexander 7
Fort Wagner, storming of (1863) **44**, 44
Fredericksburg, battle of (1862) 46

Gassendi, General 10
gauge, compound **57**, 57
Gorgas, General 45, 47, 74
Great Exhibition (1851) 55, 56
Greener 12
Greenwood, Thomas 45
Greenwood and Batley 45
Gwalior, battle of (1858) 38

Hardinge, Lord 11, 50
Hay, Colonel 75–76
Hythe, School of Musketry 50–52, **51**, **52**

Indian Mutiny (1857–58) 34–39, **35**, **36**, **37**, **38**, **39**, 54, 54, 60, 61, 74, 75
Indian service Enfields 21–22, **23**, 54
Inkerman, battle of (1854) 11, 29
Instruction of Musketry 67, **69**

Jacob, John 8
Jalalabad 36–37

Koster (gunmaker) 4

Lancaster 12, 14, 16, 22–23
Lang, Arthur Moffatt 34, 36–37, 38
le Blanc, Honoré 56
Liège-made rifles 17, 41
locks: 'hook' **12**, **13**; Pattern 1853 **13**; 'swivel' **12**, **13**
London: private arms trade 41, 46; Tower of 46, 55
Lovell 12; spring bayonet catch **15**, 15
lubricants 34, 39, 68
Lucknow, relief of (1857) 37–38, **38**; Sikander Bargh (walled garden) **37**, **38**

Macon armoury, Georgia 45
Mallet, John and Robert 55
Martini Henry rifle 23
Meerut 35, 35
Metford, William 67
Minié, Capt Claude Etienne 8, 75
Minié rifle 24, 26; bullet **11**, 11, **12**, 14, 67; Regulation Pattern 1851 **11**, 11, **12**, 14, 15
Mordecai, Maj Alfred 29, 32, 61
mortar, Mallet's 55
muskets 6, 9–10, 50; Model 1842 .69cal smoothbore 10, 46–47; rifle 17–18, 45

National Rifle Association (UK) 53
naval rifle, Pattern 1858 21 *see also* sea service rifle, Pattern 1842
Nicholas, Emperor 33
Nightingale, Florence 34

nipples 64–66, **65**

Pandey, Mangal 35, 37
Panmure, Lord 65
'pattern, sealed' 16
percussion ignition **7**, 7–8; caps, manufacture of 69–70
Porter, Adm David 48
Pritchett **14**, 14, 67
Purdey 12

Raglan, Lord 26–27, 33
ramrods 12, **62**, 62–63
Remington 57
rifle, sealed pattern, 1859 **17**, 17
rifle muskets 17–18
rifling: first use of 4–5; hexagonal bore 23; oval bore 22
Robbins and Lawrence 17, 45, 55
Royal Laboratory 67
Royal Marines 11, **25**, 25
Russell, William **28**, 28

School of Musketry 50–52, **51**, **52**
sea service rifle, Pattern 1842 11, 12 *see also* naval rifle, Pattern 1858
Sebastopol 26, 29; 'Great Redoubt' **26**, **27**; siege of (1854–55) 10, 29, **32**, 32, 33, 54, 55
Shiloh, battle of (1862) 45–46, **46**
short rifles **21**, 21, 23
sights: adjustable 'ladder' **13**, 14, 17; block rear 13; cavalry carbine **21**; smoothbored weapons for Indian service 22, **23**
Snider, Jacob 76; rifle 23, 66, 76
Springfield Armory 40; infantry rifle 15–16; musket 48–49

tactics 60–61
tallow lubricant 34, 39, 68
Thorough, Gap, battle of (1862) 48

Union Army 44, 45, 47, 75; Illinois, 20th, 39th and 101st 48; Massachusetts Infantry, 10th, 24th and 34th 49; Massachusetts Regiment, 37th 48–49; Massachusetts Volunteers, 10th 40; Massachusetts Volunteers, 20th 40–41; Massachusetts Volunteers, 54th **44**, 44, **45**; New York Volunteer Infantry, 69th ('Irish Brigade') 46; Ohio, 20th 46; Ohio Infantry, 58th 48; West Virginia, 7th 46
US Model 1855 rifled musket 45
Usherwood, Charles 29

Victoria Cross 34
Volunteer movement 52–53, 61

Waltham Watch Company 57
Waterloo, battle of (1815) 9, 28
Westley Richards 12, 13, 14
Whitworth, Joseph 8, 16, 22, 23, 53
Wilkinson 12
Windsor, Vermont 17, 55
Woolwich, Royal Arsenal 68–69, **69**

Xhosa War, Eighth (1851) 11